Advances in Motor Neuroprostheses

Ramana Vinjamuri
Editor

Advances in Motor Neuroprostheses

Springer

Editor
Ramana Vinjamuri
Stevens Institute of Technology
Hoboken, NJ, USA

ISBN 978-3-030-38739-6 ISBN 978-3-030-38740-2 (eBook)
https://doi.org/10.1007/978-3-030-38740-2

This Springer imprint is published by the registered company Springer Nature Switzerland AG.
The registered company address is: Gewerbestrasse 11, 6330 Cham, Switzerland

To my family, my teachers, and my students for their inspiration, encouragement, enthusiasm, and support.

Preface

It is an honor to be invited by Springer Publishers to edit this book titled *Advances in Motor Neuroprostheses*. The main objective of this book is to present an overview of research advances in motor neuroprostheses. A motor neuroprosthetic device is a machine that converts the intentions of the user's brain into actions. With parallel developments in neural engineering, neuroscience, deep learning, and artificial intelligence, this field of research is making great strides and bringing near natural assistive devices to market to help individuals with neuromuscular disorders.

A typical motor neuroprosthesis consists of three main modules—a signal acquisition module that obtains the neurophysiological signals, a signal processing module that filters and extracts important features from these signals, and an application module that uses these features to achieve the motion of an end effector like a prosthetic hand or a computer cursor or a wheel chair. This book includes representative examples of each of these modules and reports state-of-the-art research. Chapter "Application of Reinforcement and Deep Learning Techniques in Brain Machine Interfaces" talks about the application of reinforcement and deep learning methods in brain-machine interfaces. Chapter "Subject-Specific Muscle Activation Patterns in Athletic and Orthopedic Populations: Considerations for Using Surface Electromyography in Assistive and Biofeedback Device Applications" discusses specific muscle activation patterns in athletic and orthopedic populations in assistive and biofeedback device applications. Chapter "Kineto-Dynamic Modeling of Human Upper Limb for Robotic Manipulators and Assistive Applications" presents kineto-dynamic modeling of human upper limb for robotic manipulators and assistive applications. Chapter "Learning from the Human Hand: Force Control and Perception Using a Soft-Synergy Prosthetic Hand and Noninvasive Haptic Feedback" talks about force control and perception using a soft-synergy prosthetic hand using noninvasive haptic feedback. Chapter "Design of a Soft Glove-Based Robotic Hand Exoskeleton with Embedded Synergies" talks about the design of a soft robotic hand exoskeleton that is embedded with synergies. Chapter "Model Predictive Control Based Knee Actuator Allocation During a Standing-up Motion with a Powered Exoskeleton and Functional Electrical Stimulation" discusses model predictive control-based knee actuator allocation with a powered exoskeleton and functional electric stimulation. Chapter "Deep Brain Stimulation for Gait and Postural Disturbances in Parkinson's Disease" presents deep brain stimulation for gait and postural disturbances in Parkinson's disease. Chapter "Cognitive and Physiological Intent for the Adaptation of Motor Prostheses" talks about cognitive and physiological intent for the adaptation of motor prostheses.

With the ever-evolving research, technology, and knowledge, it is essential to provide updated information on many of these advances in motor neuroprostheses. This is the primary objective of this book. All the chapters are written by experts in this field. These chapters, and the new topics covered therein, provide a fresh perspective on our overall understanding of this area of motor neuroprostheses.

I remain thankful to all the authors who contributed their knowledge to this book. I am thankful to Springer Publishers for entrusting me with this project, and I would like to thank their entire team including Merry Stuber, Maria David, Lavanya Venkatesan, Anthony Dunlap, and Mallaigh Nolan for their support. I am indebted to my dedicated lab members and collaborators with whom I worked at different stages of production of this book. It gives me immense pleasure to present this volume to the scientific community with the hope that recent advances in our knowledge translate into a better future for individuals with disabilities who can benefit from these technologies.

Hoboken, NJ, USA Ramana Vinjamuri

Contents

Contributors

Giuseppe Averta Centro di Ricerca Enrico Piaggio, Universita di Pisa, Pisa, Italy

Xuefeng Bao Department of Biomedical Engineering, Case Western Reserve University, Cleveland, OH, USA

Gemma C. Bettelani Centro di Ricerca Enrico Piaggio, Universita di Pisa, Pisa, Italy

Matteo Bianchi Centro di Ricerca Enrico Piaggio, Universita di Pisa, Pisa, Italy

Martin K. Burns Stevens Institute of Technology, Hoboken, NJ, USA

Cosimo Della Santina Centro di Ricerca Enrico Piaggio, Universita di Pisa, Pisa, Italy

Albert Dodson Joint Department of Biomedical Engineering, North Carolina State University and University of North Carolina Chapel-Hill, Chapel Hill, NC, USA

Ehsan T. Esfahani Human-In-the-Loop System Laboratory, University at Buffalo, The State University of New York, Buffalo, NY, USA

Qiushi Fu Department of Mechanical and Aerospace Engineering, University of Central Florida, Orlando, FL, USA

David Hollinger Movement Control Rehabilitation Laboratory, Biomedical Engineering Department, Stevens Institute of Technology, Hoboken, NJ, USA

Hanyan Li Department of Biomedical Engineering, Stevens Institute of Technology, Hoboken, NJ, USA

Mingxiao Liu Movement Control Rehabilitation Laboratory, Biomedical Engineering Department, Stevens Institute of Technology, Hoboken, NJ, USA

Hemanth Manjunatha Human-In-the-Loop System Laboratory, University at Buffalo, The State University of New York, Buffalo, NY, USA

George C. McConnell Department of Biomedical Engineering, Stevens Institute of Technology, Hoboken, NJ, USA

Vahidreza Molazadeh Department of Mechanical Engineering and Materials Science, University of Pittsburgh, Pittsburgh, PA, USA

Raviraj Nataraj Movement Control Rehabilitation Laboratory, Biomedical Engineering Department, Stevens Institute of Technology, Hoboken, NJ, USA

Sean Sanford Movement Control Rehabilitation Laboratory, Biomedical Engineering Department, Stevens Institute of Technology, Hoboken, NJ, USA

Marco Santello School of Biological and Health Systems Engineering, Arizona State University, Tempe, AZ, USA

Anthony Santo Movement Control Rehabilitation Laboratory, Biomedical Engineering Department, Stevens Institute of Technology, Hoboken, NJ, USA

Nitin Sharma Joint Department of Biomedical Engineering, North Carolina State University and University of North Carolina Chapel-Hill, Chapel Hill, NC, USA

Ramana Vinjamuri Stevens Institute of Technology, Hoboken, NJ, USA

Kevin Walsh Movement Control Rehabilitation Laboratory, Biomedical Engineering Department, Stevens Institute of Technology, Hoboken, NJ, USA

Samuel Wilder Movement Control Rehabilitation Laboratory, Biomedical Engineering Department, Stevens Institute of Technology, Hoboken, NJ, USA

Antonia M. Zaferiou Department of Biomedical Engineering, Stevens Institute of Technology, Hoboken, NJ, USA

Application of Reinforcement and Deep Learning Techniques in Brain–Machine Interfaces

Hemanth Manjunatha and Ehsan T. Esfahani

Abstract From early adoption in rehabilitation, the brain–machine interfaces (BMIs) have dovetailed into applications empowering humans in controlling external devices such as prosthesis and wheelchairs with a high level of autonomy. The success of such brain–machine interfaces depends on the decoding algorithms that translate the brain activity into the human's intention or cognition state. Taking advantage of this decoding, a machine can have a robust perception of human's cognitive state and modify its actions accordingly. This decoding process can be viewed as a machine learning problem where features of brain activities are mapped to some labeled events or classes in a controlled environment. This mapping traditionally relies on the subject and task-specific signal processing approaches. Thus, the conventional machine learning methods fail to generalize well and transfer the learned features between different tasks and subjects, especially in out-of-the-lab applications. Recently, deep learning (DL) has shown great success in learning the patterns from very large data and generalizing well on different applications. With respect to brain activity analysis, deep learning and reinforcement learning (RL) techniques can significantly simplify analysis pipelines and facilitate better generalization between subjects, tasks, and also learn intricate coupled dynamics. In this regard, this paper provides information on the state of the art and challenges in implementing deep learning and reinforcement learning algorithms in brain–machine interfaces. In order to demonstrate the use of deep learning techniques in BMIs, we also present a case study of physical human–robot interaction where the brain activity is used to classify the task difficulty while interacting with the robot.

Keywords Reinforcement learning · Brain–machine interface · Motor control · EEG

H. Manjunatha · E. T. Esfahani (✉)
Human-In-the-Loop System Laboratory, University at Buffalo, The State University of New York, Buffalo, NY, USA
e-mail: hemanthm@buffalo.edu; ehsanesf@buffalo.edu

R. Vinjamuri (ed.), *Advances in Motor Neuroprostheses*,
https://doi.org/10.1007/978-3-030-38740-2_1

1 Introduction

The use of brain activity to control the external objects has become the holy grail of brain–machine interfaces. The brain activity was first recorded in 1924 by Hans Berger [1] who coined the word "electroencephalogram" (EEG) as a method for describing brain electric potentials in humans. The scalp EEG (non-invasive) picks up the summed electrical activity of postsynaptic potentials of cortical neurons and thus records the neural activity which can be used to study a wide variety of cognitive process [2].

Since EEG records electrical activity which propagates rapidly, the temporal resolution of EEG is very high. However, spatial resolution is low for non-invasive EEG as the electrical activity recorded is on the scalp. In addition, the signals that are originating in the deep brain regions are absorbed by brain tissues and skull. Hence, the signals recorded at different electrode locations in scalp EEG are highly correlated. In terms of applications, EEG (non-invasive) is widely used to study sleep pattern [3], cognitive monitoring and neuroergonomics [4–6], affective computing [7], and as a primary modality to quantify cognitive workload [8] in a variety of studies. Recently, EEG has been used as a passive approach to study human–robot interaction [9, 10], human–swarm interaction [11], virtual reality, rehabilitation [12], air traffic control, social interaction studies, and neuroprosthesis [13].

Even though EEG is emerging as an important modality in brain–machine interfaces, there are many challenges that need to be addressed. First, the EEG recordings have very low signal-to-noise ratio [14]. This is because the electrical signals originating in the deep brain regions are attenuated by the brain tissues and contaminated by electrical noises and muscle activity. Consequently, researchers use many filtering and source separation techniques to increase the signal-to-noise ratio [15]. Second, EEG signals are non-stationary and task dependent, thus making it hard to extract features that can generalize well across different tasks. Consequently, features learned in specific time windows might not generalize well towards other time instances or tasks. Even the data recorded from the same patient on different time under identical experimental conditions exhibit non-negligible difference [16]. Third, EEG recording is highly subject dependent, due to individual differences in the structure of the brain and cognitive behavior, the magnitude of response to a stimulus varies greatly. This inter-subject variability makes it difficult for classification algorithms to learn features which generalize well between subjects and tasks.

To solve the above-mentioned problems, researchers have investigated domain-specific strategies which may or may not generalize well. The typical analysis of EEG involves (1) cleaning of data from electrical line noise and removing bad quality sensor recordings. (2) Use sensor space localization algorithms such as independent component algorithms (ICA) to remove eye blink artifacts and muscle activities. (3) Obtain relevant stimuli information and segmentation of the EEG recording accordingly. (4) Obtain a baseline recording of each subject and measure the brain activity with respect to the baseline. (5) Obtain time-dependent or frequency based features such as power spectral densities and train a classification or regression model. Even though the above-mentioned steps are prevalent in EEG based studies, some of the analysis steps involve subjective assessment of the quality of operation (e.g., bad segment rejection and ICA component analysis) which makes it very hard to reproduce the studies. Hence an automated process based on the big data can help in standardizing the cleaning process and the repeatability of studies. In this context, a deep learning approach can automate pre-processing and feature extraction while generalizing well across different subjects and tasks.

2 Deep Learning in BMI Studies

The main idea of deep learning (DL) comes from an observation on the standard model of the visual cortex, which suggests that in recognizing images the brain first extracts edges, then patches, then surfaces, and finally, the complete object. This observation inspired the typical DL architecture which uses different levels of representation corresponding to characteristics or concepts hierarchies. The most prevalent architectures in DL are deep neural networks, convolution neural networks, deep belief neural networks, and recurrent neural networks. They have been successfully applied to computer vision, speech recognition, natural language processing, audio recognition, and bioinformatics [17].

Two main DL architectures that are widely used in vision and speech recognition are convolution neural networks (CNN) and recurrent neural networks (RNN) [18]. A typical CNN architecture consists of several convolution layers and fully connected layers at the end. The main idea behind CNN is learning the replicated features to detect an object/input-patch which are translated within the scene. This fundamental property achieves invariance in translation. Even scale and orientation invariance of replicated features can be achieved but at a high computational cost. Along with convolution operation pooling layers can be used to downsample the information spatially to extract the most prominent features which are influenced by the output. Figure 1 shows a typical architecture with five convolution layers and pooling operation immediately after the convolution operation. The convolution layers are succeeded by fully connected layers with a softmax activation function. For more in-depth methodology in DL techniques, the reader is referred to [18].

Regarding the application of DL techniques in EEG analysis, Yannick et al. [19] provide a scoping review of different architectures and approaches in EEG analysis. The review encompasses domains such as epilepsy, sleep, brain–computer interfacing, and cognitive and affective monitoring. They found the median gain in accuracy of DL approaches over traditional baselines to be 5.4%. However, most of the studies reported in the review were not reproducible due to lack of data and code. Along the same lines, Craik et al. [20] provide a review of DL techniques in EEG classification. They explore questions on the type of task in EEG classification, input to the neural network, and specific architecture for domain-specific applications.

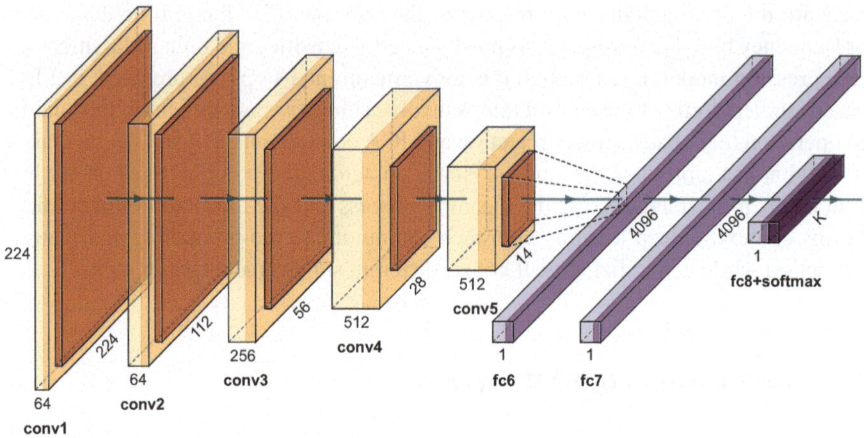

Fig. 1 A typical CNN with five convolution layers and three fully connected layers

Fig. 2 A general reinforcement learning setup

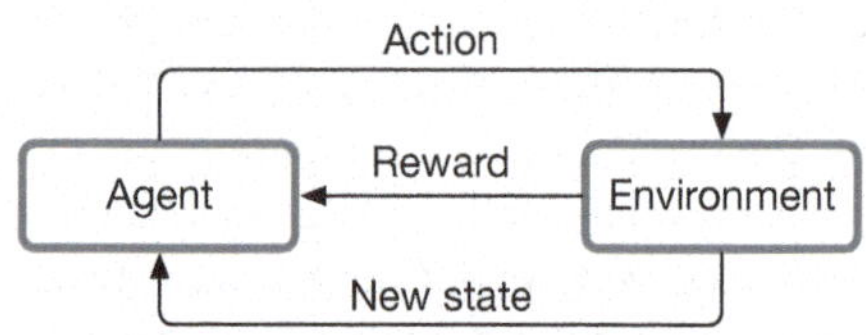

The huge amount of work reported in the above reviews bolsters the increased interest of the community in using deep learning techniques for brain–machine interfaces.

3 Reinforcement Learning in BMI Studies

Reinforcement learning is different from supervised or unsupervised learning in the sense that there is no explicit formulation of an error signal which can facilitate the learning, instead, the driving signal is replaced by a reward function [21]. Broadly speaking reinforcement learning is figuring out how to choose an action such that the reward signal over time is maximized. The reinforcement learning model can be captured in terms of Markov decision process (MDP). The main four components of reinforcement learning are an agent, environment, a reward function, and a value function. In this model, an agent acts on an environment and the environment, in turn, sends two types of signals: reward signal and a state signal (Fig. 2). The reward signal states how good the action was, whether it improved the current state, remained in the current state or the situation was worsened.

A reinforcement learning (RL) consists of a tuple of an action (a), a state (s), transition probability (Pr), reward (r), and a value function (V).

$$\text{reinforcement learning} \rightarrow \{a, s, Pr, r, V\} \tag{1}$$

At any point in time, the above-mentioned tuple should be defined for an agent. When an agent acts on an environment, it is essentially selecting an action a. As a result, its state S changes with a transition probability of Pr receiving a reward of r. The quality of action taken is evaluated by the value of V. When an agent continues to take these actions over a period of time, it gives rise to a sequence as shown in Eq. 2

$$s_0,\ a_0,\ r_0,\ s_1,\ a_1,\ r_1,\ s_2,\ a_2,\ r_2 \ldots \tag{2}$$

If this sequence ends, then the Markov decision process is said to be finite and then the tuple combination has well-defined values over that time period. In other words, given a state s and action a at time $t-1$, there exists a probability that gives the next state s' and rewards r at time t. This series of actions is known as policy (π). For learning, an agent modifies this policy in order to increase the reward over time.

There are two important methods used for learning the optimal state and action pair, namely value iteration and policy iteration. Readers are referred to [21] for more in-depth information on different reinforcement learning algorithms.

RL is particularly attractive for the brain–machine interface (BMI) applications as the agent can learn a particular set of actions in neuroprosthesis which are optimized for a specific domain. In this regard, Pohlmeyer et al. [22] used a reinforcement learning algorithm to provide an adaptive BMI controller that adapts to neural re-organizations while maintaining performance over a long period of time. Iturrate et al. [23] demonstrated an alternative BMI paradigm that overcomes decoding cognitive brain signals to achieve relevant goals for BMI. The study showed that after a short period of training, BMI operated three different neuroprosthesis and generalized across several targets. Imatz-Ojanguren et al. [24] explored the use of RL for achieving grasp functions with a surface multi-field neuroprosthesis. Prins et al. [25] used actor-critique RL model to maximize the performance of the BMI. Their study showed that the training time in BMI was significantly reduced. These are few works relating RL and brain–machine interfaces that can drastically change the neural-decoding procedure in neuroprosthesis field.

4 Application of Deep Learning in Human–Robot Interaction: A Case Study

In order to demonstrate the use of deep learning techniques for brain–machine interface applications, we present a case study of physical human–robot interaction where the task difficulty of interaction is modified while brain activity is recorded simultaneously. A convolution neural network is used to classify the task difficulty from the brain activity using two different architectures and pre-processing steps. This case study shows the importance of choosing the appropriate network architecture and pre-processing steps which depends on the domain-specific problems.

In an experiment approved by the Institutional Review Board, 11 subjects (8 males and 3 females; all right-handed) were recruited from the University of Buffalo School of Engineering. Participants' age ranged from 23 to 34 years (Mean = 24, SD = 2.1). The participants were asked to hold the end effector of a robotic manipulator connected to a force-torque sensor (Fig. 3) and guide it through a star-shaped pathway as shown in Fig. 4. An admittance controller was used to regulate the position of the end effector based on the applied force. The dynamics of the interaction was changed by altering the damping in the admittance control (all the other parameters were kept constant). Participants traced the trajectory with low damping (high admittance) and high damping (low admittance) in separate trials. In low damping condition, the resistance offered by the robot is less and thus the effort by the human to move the end effector is relatively low, conversely, in high damping as the resistance is high, effort requirement is also high.

The relation between the force applied (F_h) and velocity of the end effector is given as the first linear differential equation as shown in Eq. 3

$$M_d\dot{v} + C_d v = F_h \tag{3}$$

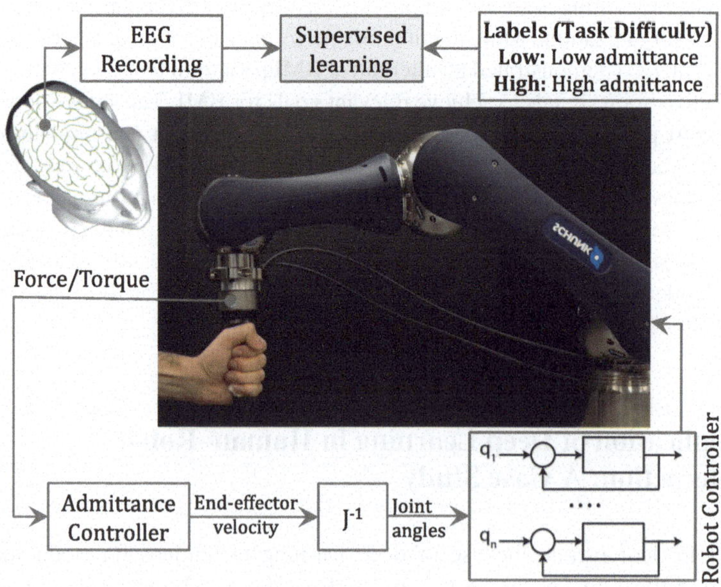

Fig. 3 Physical human–robot interaction experiment setup and brain activity recording

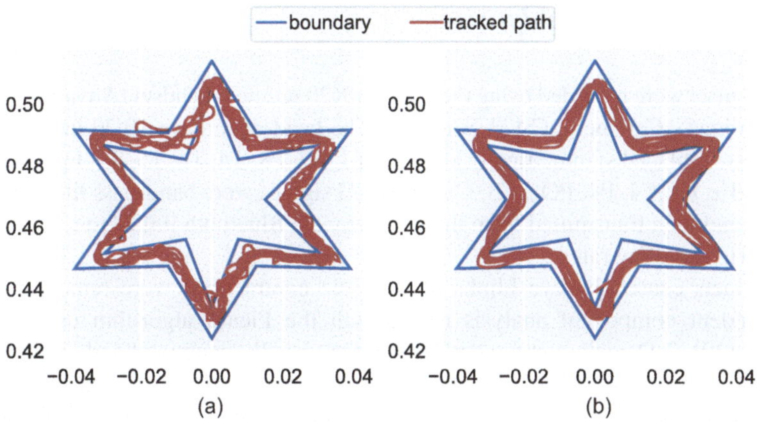

Fig. 4 Tracked path and boundary for (**a**) high admittance and (**b**) low admittance conditions

where M_d and C_d are the mass and damping coefficient matrices for each joint. The values for M_d and C_d are taken as $[3, 3, 3, 0.1, 0.1, 0.1]$ and $[20, 20, 20, 6, 6, 6]$, respectively. This setting corresponds to low admittance of the robot, i.e., the resistance offered by the robot while the human is applying the force is less. Equation 3 is integrated in order to obtain the velocity of the end effector. Then Eq. 4 is used to get the joint velocities and torques. However, the solution is not unique for a given set of end effector position. For more information, the reader is referred to [26].

$$\dot{q} = J^{-1}(q)v_c, \; \tau = J^T(q)F_h \tag{4}$$

Under high admittance (low damping), the robot becomes highly sensitive to the change in the input force at the end effector (in this experiment, the input force is from human hand). Under low admittance as the damping is high, the robot is more stable. Thus, under high admittance, it is more difficult for a human to control the robot motion due to unstable motion. Conversely, under low admittance, the robot motion is smooth and controlled as the damping is high. This is evident as shown in Fig. 4. The movements under high admittance have more disturbances (Fig. 4a) when compared to movements in low admittance (Fig. 4b). The classification algorithm is trained to distinguish between these two conditions. For the classification procedure, high admittance condition was chosen as first class and low admittance condition was chosen as second class. The brain activity recorded during these two conditions was used as the input for the deep learning algorithm.

4.1 EEG Signal Analysis

EEG signals were recorded using the B-Alert X20 wireless headset (Advanced Brain Monitoring©. Carlsberg, CA, USA) [27]. The headset follows 10/20 international [28] system placed symmetrically at sites: Fp1, F7, F8, T4, T6, T5, T3, Fp2, O1, P3, Pz, F3, Fz, F4, C4, P4, POz, C3, Cz, O2. EEG signals were band-pass filtered (0.1–70 Hz) and then transmitted from the headset via a Bluetooth linked to a nearby PC at 256 Hz sampling rate.

Artifacts caused by eye blinks and muscle contractions were removed using independent component analysis (ICA) with the Picard algorithm using MNE python [29]. 2-D scalp component maps were visually examined and components corresponding to eye movements and non-cognitive activities were removed. After removal, the components were projected back to get artifact-free EEG signal (Fig. 5).

Furthermore, a Hamming window with 50% overlap was used to extract relative and absolute power spectral densities using the Welch method from 1-s epochs. Features were extracted from three frequency bands, namely Theta (4–7 Hz), Alpha (8–13 Hz), and Beta (14–35 Hz).

4.2 Convolution Neural Network Architecture

Figure 6 shows the architecture of the convolution neural network used for the classification of task difficulty. The network has three convolution layers receiving

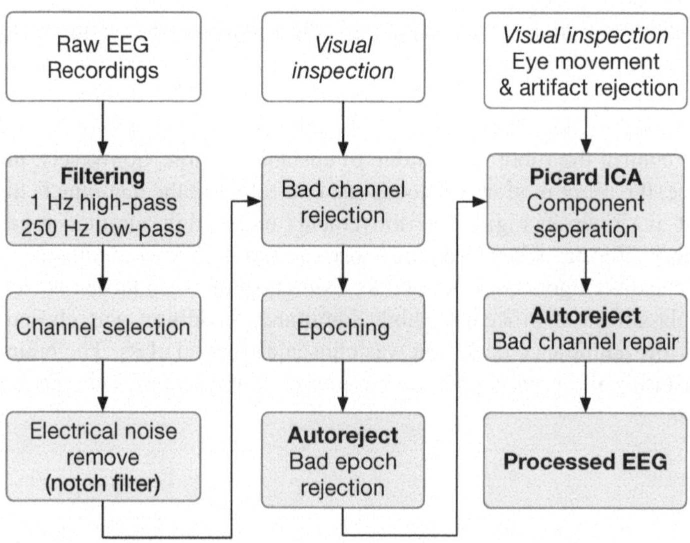

Fig. 5 EEG analysis pipeline

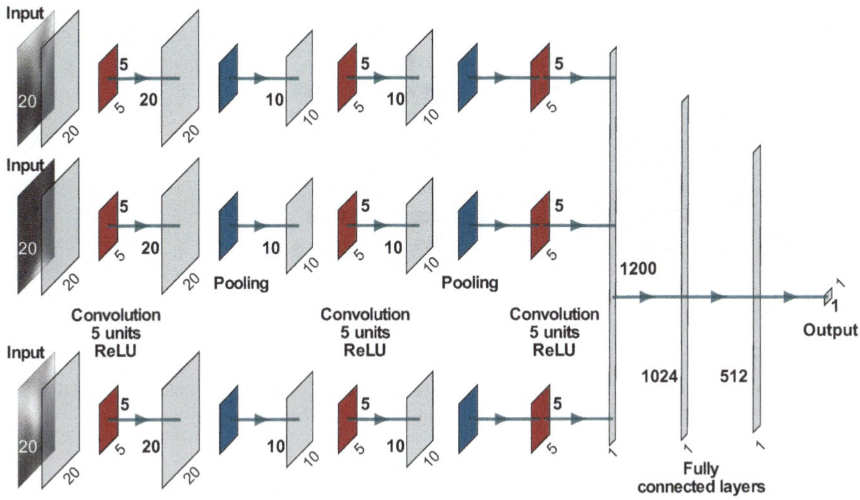

Fig. 6 Convolution neural network architecture 1 for task difficulty prediction

three images corresponding to Theta, Alpha, and Beta bands. Each convolution layer consists of two convolution operation with exponential linear unit (ELU) activation and Max pooling in series (Fig. 6). The output from each convolution layer is flattened and stacked with output from other convolution layers to form flattened features layer. These features are used as inputs for two fully connected layers with a rectified linear unit (ReLU) activation and dropout (75% keep probability). For loss function, the log-softmax is used.

For the architecture shown in Fig. 6, the EEG was prepossessed to form the spectral images and coherence images instead of using the EEG signals directly. With the cleaned EEG, topology images were constructed by first mapping the sensor space into a grid of required dimension (here it is the size of the image, 28 by 28). The mapping is done by constructing a grid of required x (28) and y (28) dimension and assign the coordinates of the grid to coordinates of the EEG electrodes (sensor locations). The values of power spectral density at these locations are calculated using the Welch method. The values of power spectral density (and coherence) at remaining grid positions are calculated using bi-linear interpolation. This procedure produces the image of (Fig. 7) required dimension and values of power spectral density (Fig. 7a, PSD) and coherence (COH) which are used as inputs for the convolution neural networks

For constructing coherence image, there are two options: one to convert the coherence to an image where each column corresponds to an EEG electrode (Fig. 7b). Second, construct an interpolated images same of PSD images. In this study, we have used the interpolation procedure to construct the image (Fig. 7c) of the coherence and the same has been used in convolution neural network. The

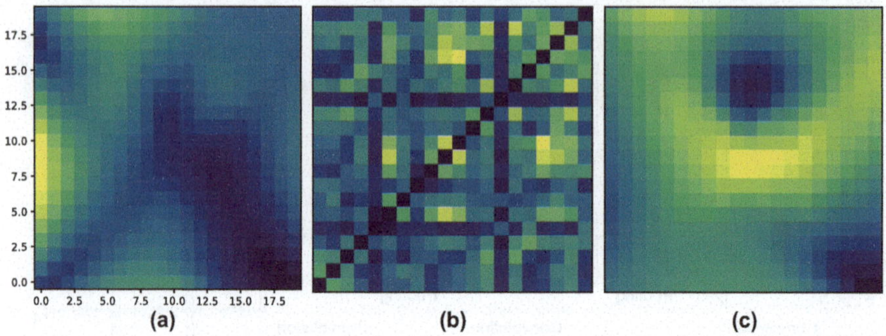

Fig. 7 (**a**) Power spectral density of EEG reading as an image. (**b**) The coherence among different EEG electrodes represented as connectivity matrix. (**c**) Same coherence values represented as an image (all images show were constructed for alpha 11–13 Hz EEG recording)

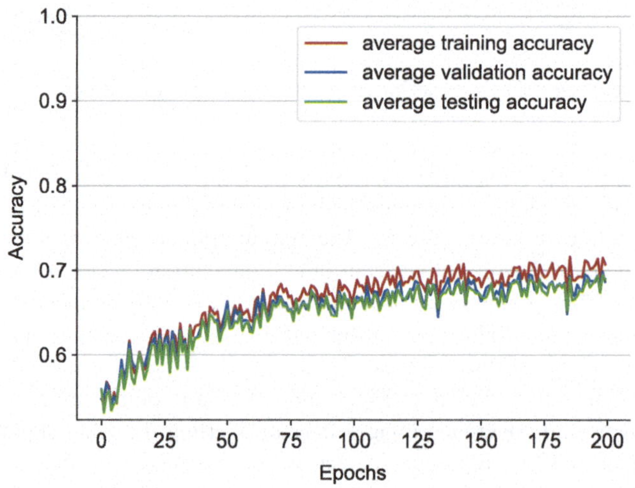

Fig. 8 Classification accuracy of training, validation, and testing data when architecture 1 CNN is used

convolution neural network is trained on pooled data of all the subjects (70% training, 15% validation, and 15% testing) using the PyTorch framework.

Figure 8 shows the training, validation, and testing accuracy for 200 epochs. As can be seen, the accuracy is not very high with respect to a chance (50%, two classes). Nonetheless, there is no over-fitting as validation accuracy is very close to training accuracy. Increasing the number of epochs and learning rate may improve the classification accuracy.

Figure 9 shows the convolution neural network architecture adapted from [30]. The network consists of four convolution operations with a log-softmax output and no fully connected layers. A raw EEG signal of a 2-s epoch is used as an input to the neural network and the first convolution operation is carried across time and

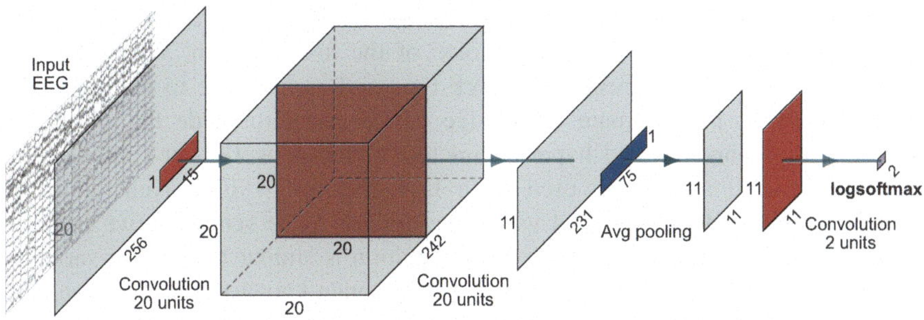

Fig. 9 Convolution neural network architecture 2 for task difficulty prediction in HRI

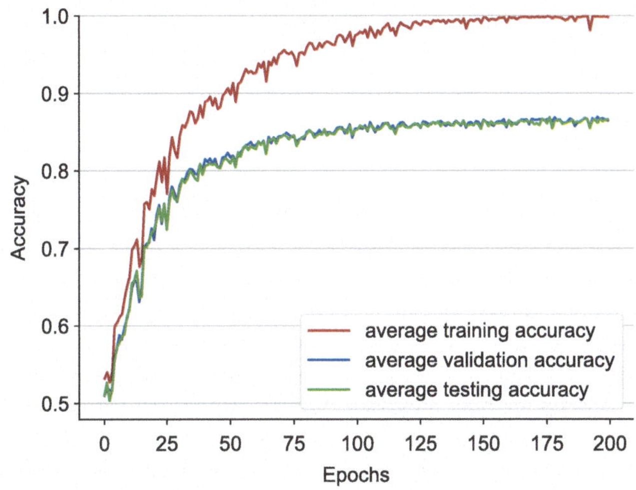

Fig. 10 Classification accuracy of training, validation, and testing data when architecture 2 CNN is used

thus it captures temporal information. The second convolution operation is carried out across the space capturing spatial information. After the temporal and spatial convolution operation, the output is squared and log transformed. The resulting output is convoluted with the remaining two convolution layers sequentially to produce the class label output.

The temporal and spatial convolution can be combined into one 3D convolution operation; however, the choice of splitting the 3D convolution into two 2D convolutions facilitates the study of learned features in time and spatial domain.

Figure 10 shows the accuracy of classification using the second architecture. The accuracy is significantly higher than the classification accuracy in the first architecture. However, the network slightly overfits towards the end of the training. This can be addressed by using dropout in the last convolution layers [31].

This case study shows the importance of the choice of neural network architecture, pre-processing of data, and choice of the loss function. While there is no systematic way of addressing network type and architecture to use for brain activity analysis, prior-domain knowledge can be used to guide the choice of deep neural architecture and hyper-parameters of the network. For example, the second architecture that was used is in fact inspired by filter bank common spatial patterns (FBCSP) method which has proven to be very effective in EEG classification. The squaring and log transformation is similar to trial log-variance computation in FBCSP [32]. Incorporating the domain knowledge in the design of architecture allows the use of a single network to decode the brain activity and learn relevant features from the data. The use of deep learning techniques for brain data decoding opens up a lot of avenues in neuroprosthesis, brain–computer interfaces, psychological studies, and affective computing.

5 Conclusion

The brain–machine interfaces are poised towards empowering humans in controlling external devices such as prosthesis and wheelchairs. For the success of such technologies, decoding of brain signals is imperative. Many researchers have developed domain and task-specific brain decoding methods which are uphill while adapting to other domains. Recently, deep learning has shown great success in learning the patterns from very large data while generalizing well across different applications. A deep learning approach through end-to-end learning can automate pre-processing, feature extraction, and brain-signal decoding. To demonstrate such an approach, a case study is presented where the task difficulty during physical human–robot interaction was predicted from the brain activity (through EEG) using deep learning techniques. Two architectures involving convolution neural networks are presented, with the first architecture using processed and transformed EEG data and the second architecture using raw EEG data. The case study demonstrates that the choice of architecture is very important in achieving better performance. Even though there is no systematic way of selecting network type and architecture, prior-domain knowledge can be used to guide the choice of deep neural architecture and hyper-parameters of the network. In spite of countable shortcomings such as the need for huge data and domain-specific network architecture, deep learning has a lot of potential in streamlining brain-signal decoding procedure in brain–machine interfaces.

Acknowledgment We gratefully acknowledge the support of NVIDIA Corporation with the donation of the Titan Xp GPU used for this research.

References

1. M. Teplan et al., Fundamentals of EEG measurement. Meas. Sci. Rev. **2**(2), 1–11 (2002)
2. R.P.N. Rao, *Brain-Computer Interfacing: An Introduction* (Cambridge University Press, Cambridge, 2013)
3. S. Motamedi-Fakhr, M. Moshrefi-Torbati, M. Hill, C.M. Hill, P.R. White, Signal processing techniques applied to human sleep EEG signals—a review. Biomed. Signal Process. Control **10**, 21–33 (2014)
4. K.A. Guru, E.T. Esfahani, S.J. Raza, R. Bhat, K. Wang, Y. Hammond, G. Wilding, J.O. Peabody, A.J. Chowriappa, Cognitive skills assessment during robot-assisted surgery: separating the wheat from the chaff. BJU Int. **115**(1), 166–174 (2015)
5. A.H. Memar, E.T. Esfahani, Physiological measures for human performance analysis in human-robot teamwork: case of tele-exploration. IEEE Access **6**, 3694–3705 (2018)
6. M. Rahman, W. Karwowski, M. Fafrowicz, P.A. Hancock, Neuroergonomics applications of electroencephalography in physical activities: a systematic review. Front. Hum. Neurosci. **13**, 182 (2019)
7. M.-K. Kim, M. Kim, E. Oh, S.-P. Kim, A review on the computational methods for emotional state estimation from the human EEG. Comput. Math. Methods Med. **2013**, 13 pp. (2013)
8. P. Zarjam, J. Epps, N.H. Lovell, Beyond subjective self-rating: EEG signal classification of cognitive workload. IEEE Trans. Auton. Ment. Dev. **7**(4), 301–310 (2015)
9. E.T. Esfahani, V. Sundararajan, Using brain-computer interfaces to detect human satisfaction in human-robot interaction. Int. J. Humanoid Rob. **08**(01), 87–101 (2011)
10. X. Mao, W. Li, C. Lei, J. Jin, F. Duan, S. Chen, A brain–robot interaction system by fusing human and machine intelligence. IEEE Trans. Neural Syst. Rehabil. Eng. **27**(3), 533–542 (2019)
11. G.K. Karavas, D.T. Larsson, P. Artemiadis, A hybrid BMI for control of robotic swarms: preliminary results, in *2017 IEEE/RSJ International Conference on Intelligent Robots and Systems (IROS)* (IEEE, 2017), pp. 5065–5075
12. K.K. Ang, K.S.G. Chua, K.S. Phua, C. Wang, Z.Y. Chin, C.W.K. Kuah, W. Low, C. Guan, A randomized controlled trial of EEG-based motor imagery brain-computer interface robotic rehabilitation for stroke. Clin. EEG Neurosci. **46**(4), 310–320 (2015)
13. P. Ofner, A. Schwarz, J. Pereira, G. Müller-Putz, Decoding movements of the upper limb from EEG, in *cuttingEEG* (2017)
14. N. Bigdely-Shamlo, T. Mullen, C. Kothe, K.-M. Su, K.A. Robbins, The PREP pipeline: standardized preprocessing for large-scale EEG analysis. Front. Neuroinform. **9**, 16 (2015)
15. W. Zhang, C. Tan, F. Sun, H. Wu, B. Zhang, A review of EEG-based brain-computer interface systems design. Brain Sci. Adv. **4**(2), 156–167 (2018)
16. M.X. Cohen, *Analyzing Neural Time Series Data: Theory and Practice* (MIT Press, Cambridge, 2014)
17. M.M. Najafabadi, F. Villanustre, T.M. Khoshgoftaar, N. Seliya, R. Wald, E. Muharemagic, Deep learning applications and challenges in big data analytics. J. Big Data **2**(1), 1 (2015)
18. I. Goodfellow, Y. Bengio, A. Courville, *Deep Learning* (MIT Press, Cambridge, 2016)
19. R. Yannick, B. Hubert, A. Isabela, G. Alexandre, F. Jocelyn et al., Deep learning-based electroencephalography analysis: a systematic review (2019). Preprint. arXiv:1901.05498
20. A. Craik, Y. He, J.L. Contreras-Vidal, Deep learning for electroencephalogram (EEG) classification tasks: a review. J. Neural Eng. **16**(3), 031001 (2019)
21. R.S. Sutton, A.G, Barto, *Reinforcement Learning: An Introduction* (MIT Press, Cambridge, 2018)
22. E.A. Pohlmeyer, B. Mahmoudi, S. Geng, N.W. Prins, J.C. Sanchez, Using reinforcement learning to provide stable brain-machine interface control despite neural input reorganization. PloS One **9**(1), e87253 (2014)
23. I. Iturrate, R. Chavarriaga, L. Montesano, J. Minguez, J.d.R. Millán, Teaching brain-machine interfaces as an alternative paradigm to neuroprosthetics control. Sci. Rep. **5**, 13893 (2015)

24. E. Imatz-Ojanguren, E. Irigoyen, T. Keller, Reinforcement learning for hand grasp with surface multi-field neuroprostheses, in *International Joint Conference* (Springer, Berlin, 2016), pp. 313–322
25. N.W. Prins, J.C. Sanchez, A. Prasad, Feedback for reinforcement learning based brain–machine interfaces using confidence metrics. J. Neural Eng. **14**(3), 036016 (2017)
26. A.H. Memar, E.T. Esfahani, EEG correlates of motor control difficulty in physical human-robot interaction: a frequency domain analysis, in *2018 IEEE Haptics Symposium (HAPTICS)* (IEEE, 2018), pp. 229–234
27. C. Berka, D.J. Levendowski, M.N. Lumicao, A. Yau, G. Davis, V.T. Zivkovic, R.E. Olmstead, P.D. Tremoulet, P.L. Craven, EEG correlates of task engagement and mental workload in vigilance, learning, and memory tasks. Aviat. Space Environ. Med. **78**(5), B231–B244 (2007)
28. R.W. Homan, J. Herman, P. Purdy, Cerebral location of international 10-20 system electrode placement. Electroencephalogr. Clin. Neurophysiol. **66**(4), 376–382 (1987)
29. A. Gramfort, M. Luessi, E. Larson, D.A. Engemann, D. Strohmeier, C. Brodbeck, R. Goj, M. Jas, T. Brooks, L. Parkkonen et al., MEG and EEG data analysis with MNE-python. Front. Neurosci. **7**, 267 (2013)
30. R.T. Schirrmeister, J.T. Springenberg, L.D.J. Fiederer, M. Glasstetter, K. Eggensperger, M. Tangermann, F. Hutter, W. Burgard, T. Ball, Deep learning with convolutional neural networks for EEG decoding and visualization. Hum. Brain Mapp. **38**(11), 5391–5420 (2017)
31. N. Srivastava, G. Hinton, A. Krizhevsky, I. Sutskever, R. Salakhutdinov, Dropout: a simple way to prevent neural networks from overfitting. J. Mach. Learn. Res. **15**(1), 1929–1958 (2014)
32. K.K. Ang, Z.Y. Chin, H. Zhang, C. Guan, Filter bank common spatial pattern (FBCSP) in brain-computer interface, in *2008 IEEE International Joint Conference on Neural Networks (IEEE World Congress on Computational Intelligence)* (IEEE, 2008), pp. 2390–2397

Subject-Specific Muscle Activation Patterns in Athletic and Orthopedic Populations: Considerations for Using Surface Electromyography in Assistive and Biofeedback Device Applications

Antonia M. Zaferiou

Abstract This chapter presents the subject-specific nature of muscle activation patterns measured by surface electromyography (sEMG) both when elite athletes perform complicated whole-body maneuvers and when patients perform activities of daily living. Examples are provided to highlight the vast differences in muscle activation patterns across two ballet dancers, baseball pitchers ($n = 16$), and preoperative reverse total shoulder arthroplasty patients ($n = 6$). These subject-specific muscle activations correspond to subject-specific movement mechanics and are relevant to consider while designing and testing devices that use sEMG signals as inputs. This "sample course" suggests that thresholds, normalization, and filtering of electromyography signals as inputs to assistive or biofeedback devices need to be carefully selected per individual. Recently, there have been exciting advances in machine learning (Campopiano et al., Behav. Brain Res. **347**:425–435, 2018) and electromyography technology (i.e., multi-node sEMG arrays (Farina et al., J. Appl. Physiol. **117**:1215–1230, 2018)) that may assist in personalizing devices, more robustly normalizing signals, and/or identifying control commands.

Keywords Electromyography · Subject-specific · Personalize · Sports · Orthopedic · Biomechanics

1 Muscle Activation Measurement and Analysis

Dual surface electrodes with an interelectrode distance of 1 cm (Noraxon, Scottsdale, AZ) were placed along the fibers and over each muscle belly. Muscle activation data were filtered using a fourth-order butterworth zero-phase band-pass filter (10–400 Hz or 40–400 Hz on trunk sEMG signals that included cardiac signal) using

A. M. Zaferiou (✉)
Department of Biomedical Engineering, Stevens Institute of Technology, Hoboken, NJ, USA
e-mail: Antonia.Zaferiou@stevens.edu

R. Vinjamuri (ed.), *Advances in Motor Neuroprostheses*,
https://doi.org/10.1007/978-3-030-38740-2_2

custom MATLAB software (Mathworks, Novi, MI). Signals were then rectified and placed into 20 ms average “bins” [1–3].

All examples provided used the maximum binned sEMG signals during isometric manual muscle tests (MMT) or modified MMTs in clinical populations in order to normalize the sEMG measured during movement tasks of interest [4]. Baseball and shoulder arthroplasty examples present muscle activation color “heatmaps” relative to a within-task maximum for that muscle for that individual or relative to a stated percentage of the maximum bin during that muscle’s MMT trial. All heatmaps display the activation level of muscles such that dark red represents high activity and dark blue represents low activity.

2 Dancers Using Subject-Specific Muscle Activation Patterns During Turns

The muscle activation patterns of two exemplar dancers during the turn initiation phase are plotted in Fig. 1 to compare single- to double-revolution piqué turns and in Fig. 2 to compare double- to triple-revolution piqué turns. During the turn initiation phase, the body starts stationary in single-limb stance with the rear (or “push”) leg

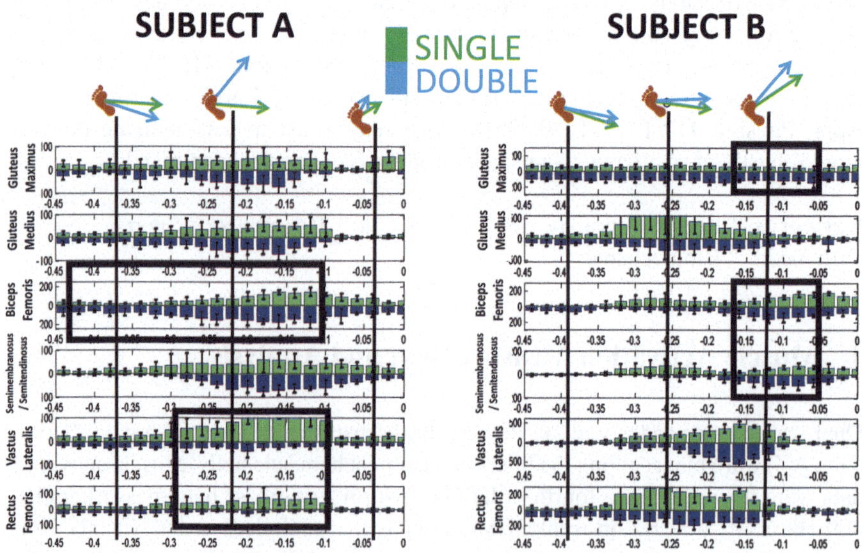

Fig. 1 Exemplar mean (SD) binned muscle activation vs. time of the rear leg during the turn initiation phase of single (positive, green) vs. double (negated, blue) piqué turns, as a percent max MMT for Subject A and Subject B. Multiple trials were synced at time zero, which is the start of the turn phase when the rear leg departs from the ground. Above each graph, cartoons depict the horizontal ground reaction force context for each subject at three times during the turn initiation phase (green vector for single turn, blue vector for double turn)

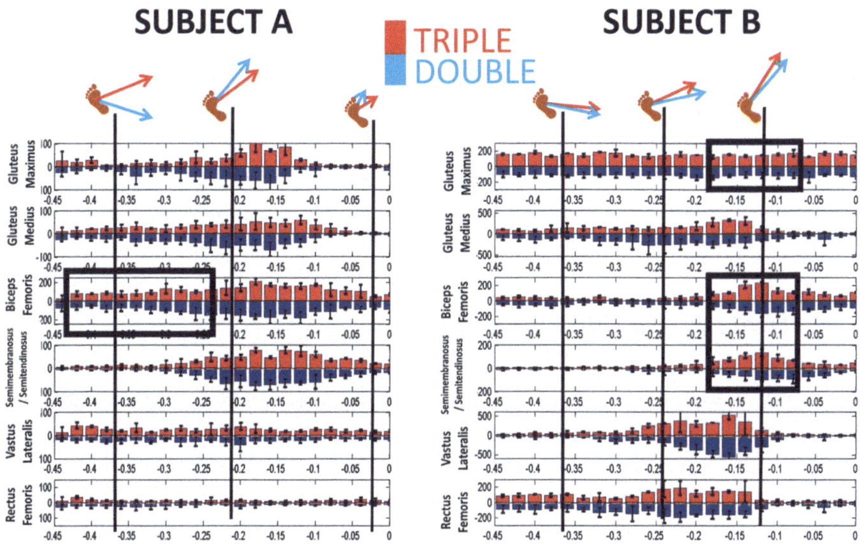

Fig. 2 Mean (SD) binned muscle activation vs. time of the rear leg during double (positive, blue) vs. triple (negated, red) piqué turns, as a percent max MMT for Subject A and Subject B. Multiple trials were synced at time zero, which is the start of the turn phase when the rear leg departs from the ground. Above each graph, cartoons depict the horizontal ground reaction force context for each subject at three times during the turn initiation phase (blue vector for double turn, red vector for triple turn)

in contact with the ground, as previously described by Zaferiou et al. [5]. Both dancers used distinct horizontal ground reaction force patterns and distinct muscle activation patterns for all the three turn types. When the foot is in contact with the ground, lower extremity muscle activation patterns can elicit distinct ground reaction force patterns. For example, when seated on a rolling chair with the feet on the ground, activation of the hamstrings can cause the foot to push on the ground posteriorly (backwards), which elicits an anterior (forward) ground reaction force that would accelerate the center of mass anteriorly (rolling the chair forward). Using an understanding of this relationship, the muscle activation patterns observed in these dancers are consistent with the measured ground reaction force patterns (Figs. 1 and 2). Further information about the piqué turn and the experimental setup can be found in Zaferiou et al. [5].

2.1 Double vs. Single Piqué Turn

Subject A increases the activation of her biceps femoris in a double vs. single turn, which coincides with increasing the anterior reaction forces in a double vs. single turn (Fig. 1). Subject A also decreases the activation of the quadriceps (vastus lateralis and rectus femoris), which is consistent with decreased medial-posteriorly directed reaction forces (because the leg is externally rotated) (Fig. 1). Subject B, in contrast, increases the gluteus maximus, hamstring (biceps femoris and semimembranosus/semitendinosus) activation in a double vs. single turn, which coincides with increasing the anterior ground reaction force (because the leg is externally rotated) in a double vs. single turn (Fig. 1), without consistently decreasing the activation of her quadriceps (large standard deviation) as did Subject A.

2.2 *Triple vs. Double Piqué Turns*

Subject A increases the activation of her biceps femoris (earlier onset of activation) in a triple vs. double turn, which coincides with increasing the anterior reaction forces in a triple vs. double turn earlier (Fig. 2). Subject B, in contrast, increases the gluteus maximus (slight increase of gluteus medius) and hamstrings (biceps femoris and semimembranosus/semitendinosus) activation in a triple vs. double turn, which coincides with increasing the anterior ground reaction force (because the leg is externally rotated) in a double vs. single turn (Fig. 2).

2.3 *Take-Homes from This Exemplar Comparison*

This small yet detailed comparison is exemplar of the subject-specific mechanics used by a larger group of dancers performing piqué, classical pirouette turns, and contemporary/modern pirouette turns of varied rotational demands while achieving similar whole-body kinematics [5–7]. With this exemplar comparison, the within-subject consistency of muscle activation can be appreciated (distinct "on/off" patterns and activation "shapes" with relatively small standard deviation bars). Note, the synchronization event trial-to-trial was at the end of the initiation phase, which could present practical challenges to online use of EMG as inputs to control devices (vs. using an event at the beginning).

From a control perspective, Subject A seems to simplify control by modulating the activation of one key muscle (biceps femoris) as rotation demand increases, whereas Subject B can be seen to simplify control by using a consistent strategy to change in muscle activation (scaling activation of the same three muscles) from single to double to triple turns. Additionally, from a physical therapy perspective, the way that Subject B increases activation of gluteus maximus with increased hamstring activation may provide protective stability to the hip joint against anterior femoral slide [8].

3 Baseball Pitchers Using Subject-Specific Hamstring Muscle Recruitment

Figure 3 displays a muscle activation heatmap for four hamstrings (bilateral biceps femoris and semimembranosus/semitendinosus) while elite baseball players pitched (Sub 1–Sub 16). All trials were synchronized to back leg's departure from the floor, at time zero. In this example, there is a notable variance between individuals in muscle activation patterns. However, future pitching studies will include full-body motion capture to provide further context for these muscle activation patterns. Further information about the activation patterns of these hamstrings and the implications for orthopedic decision-making regarding harvesting grafts from these muscles for UCL repair is described by Erickson et al. [9]. Note, the synchronization event trial-to-trial was when the back leg lifted from the ground, which could present practical challenges to online use of EMG as inputs to control devices (vs. using an event at the beginning).

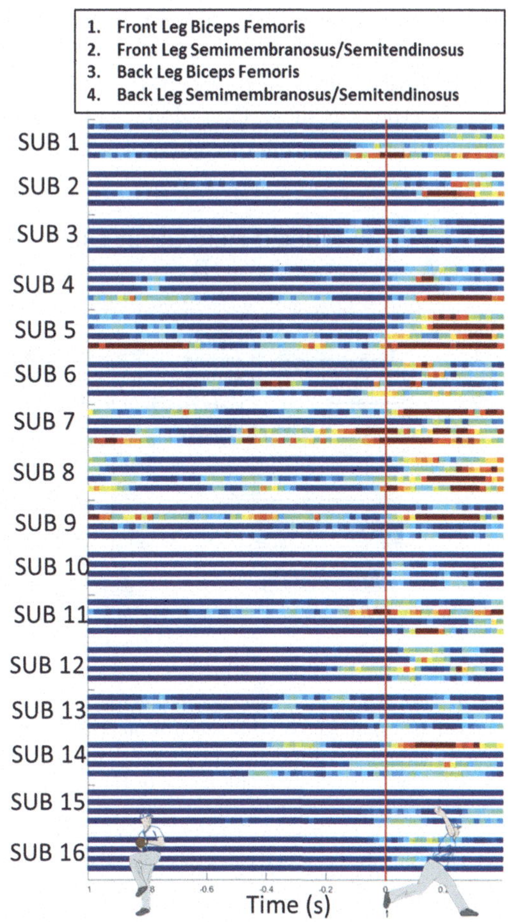

Fig. 3 Muscle activation heatmap during the baseball pitch for 16 pitchers: color is normalized to 50% MMT such that dark red represents when the muscle is active to at least 50% of the maximum 20 ms bin measured during the MMT trial. The graphs start when the front leg lifts from the ground and time zero (red vertical line) is when the back leg departs from the ground

4 Preoperative Shoulder Arthroplasty Patients Using Subject-Specific Movement Mechanics During Arm Elevation Tasks

Figures 4 and 5 display muscle activation heatmaps during simulated brushing teeth (Fig. 4) and scratching the lower back (Fig. 5) performed by six preoperative reverse total shoulder arthroplasty patients. Patients used different muscle activation patterns during this task along with different kinematics. However, when the color is normalized to "within-task max," in Fig. 4b, it is notable that there are similarities across patients, such as maximum activation for most muscles occurs as the arm reaches peak elevation (near time zero) or soon after, as the hand simulates brushing teeth. During scratching the low back in Fig. 5, we see different muscle activation patterns vs. Fig. 4 consistent with the difference in kinematics between tasks. For example, compared to brushing teeth (Fig. 4a, b), during scratching the low back (Fig. 5a, b), most patients used relatively larger pectoralis major activation as the arm oscillated with the hand behind the low back (more orange) and less anterior deltoid (more blue) consistent with the kinematic context differences. Comparison between Figs. 4 and 5 parts a, b, and c demonstrates how sensitive interpretation of the sEMG signal is to decisions about how to normalize and express sEMG data. Interpreting these data along with the kinematic or kinetic contexts will provide more robust interpretation of subject-specific movement mechanics. Note,

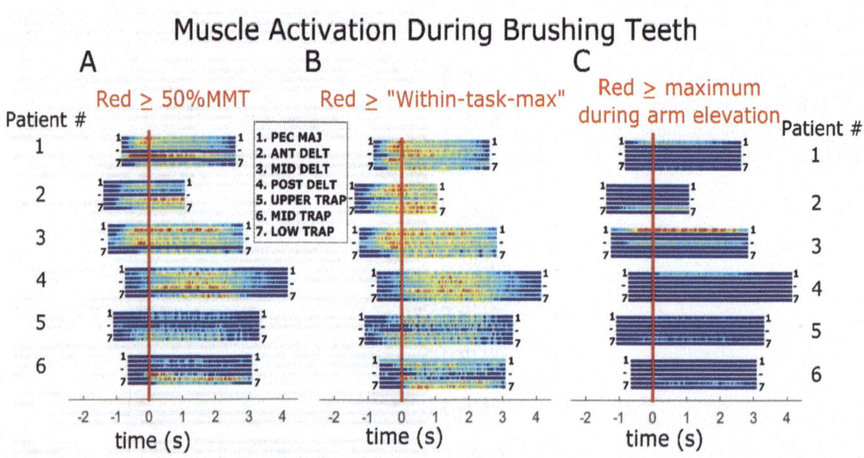

Fig. 4 Muscle activation heatmaps during trials in which patients 1–6 elevated their preoperative arm and then simulated brushing their teeth. These muscle activation patterns are expressed with different color normalizations such that dark red indicates that the muscle was active to at least (**a**) 50% MMT, (**b**) "within-task max," or (**c**) the maximum activation bin during arm elevation trials, whereas dark blue indicates low activation levels. Each color "pixel" was determined first by syncing all trials to time zero, when the elbow reaches its maximum height before teeth brushing, and then using the mean activation during a 20 ms bin across at least three repetitions

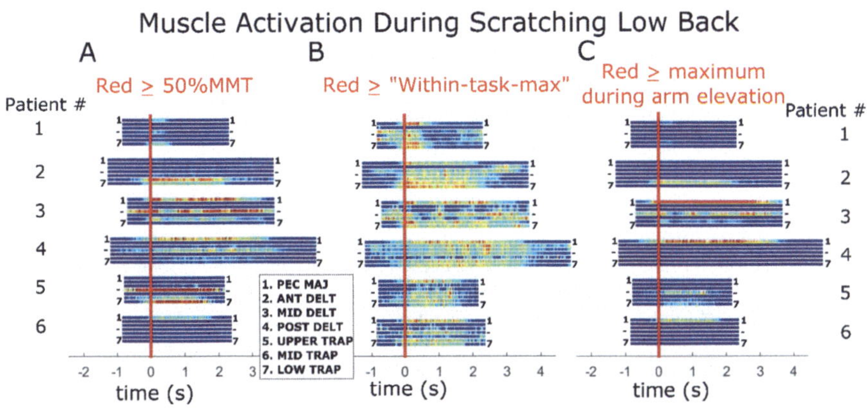

Fig. 5 Muscle activation heatmaps during trials in which patients 1–6 scratched their low back. These muscle activation patterns are expressed with different color normalizations such that dark red indicates that the muscle was active to at least (**a**) 50% MMT, (**b**) "within-task max", or (**c**) the maximum activation bin during arm elevation trials, whereas dark blue indicates low activation levels. Each color "pixel" was determined first by syncing all trials to time zero, when the elbow reaches its most posterior position before scratching began, and then using the mean activation during a 20 ms bin across at least three repetitions

the synchronization event trial-to-trial was when the elbow reached its maximum height, which could present practical challenges to online use of EMG as inputs to control devices (vs. using an event at the beginning).

5 Concluding Remarks

This chapter presented examples of subject-specific muscle activation patterns measured by sEMG when athletes and patients perform a variety of tasks. These subject-specific muscle activations correspond to subject-specific movement mechanics and are relevant to consider while designing and testing devices that use sEMG signals as inputs. When the kinematic and kinetic contexts are known, sEMG data become more useful to interpretation or use in devices. For instance, future work with the shoulder arthroplasty dataset will include analyzing muscle activation patterns vs. arm elevation angle (varied across patients) and vs. net joint kinetics. Using these frameworks may elucidate muscle activation patterns that may be shared across all patients, or a subset of patients. This chapter also demonstrated that people may have similar performance outcomes (e.g., the two dancers in Figs. 1 and 2) but use different strategies, including muscle activation patterns, to accomplish similar whole-body dynamics.

Careful considerations should be taken when deciding thresholds, normalization, and filtering of sEMG signals if they are to be used as inputs to assistive or biofeedback devices. Often the most effective synchronization event trial-to-trial will not be at the beginning of the movement of interest, which could present practical challenges to online use of EMG as inputs to control devices. Personalizing assistive or biofeedback approaches will advance the field of precision medicine and enable elite athletes to improve their movement mechanics iteratively, after first appreciating how the person initially controls his/her movement.

Acknowledgments This work was made possible by volunteer research participants, research assistants, and various funding sources. The dance study was supported by the University of Southern California Biomechanics Research Laboratory and the University of Southern California Myronis Endowed Fellowship. Baseball and clinical shoulder studies were funded by Rush University Medical Center and philanthropic donations to the Department of Orthopedic Surgery at Rush University Medical Center. The clinical shoulder work was also supported by the Cohn Fellowship, administered through Faculty Affairs at Rush University.

References

1. A. Campopiano, S.J.R. van Noordt, S.J. Segalowitz, STATSLAB: an open-source EEG toolbox for computing single-subject effects using robust statistics. Behav. Brain Res. **347**, 425–435 (2018)
2. D. Farina, R. Merletti, R.M. Enoka, The extraction of neural strategies from the surface EMG: an update. J. Appl. Physiol. **117**, 1215–1230 (2018). https://doi.org/10.1152/japplphysiol.00162.2014
3. C.J. De Luca et al., Filtering the surface EMG signal: Movement artifact and baseline noise contamination. J. Biomech. **43**, 1573–1579 (2010)
4. F.P. Kendall et al., *Muscles: Testing and Function, with Posture and Pain* (Lippincott Williams & Wilkins, Baltimore, 1993)
5. A.M. Zaferiou et al., Modification of impulse generation during piqué turns with increased rotational demands. Hum. Mov. Sci. **47**, 220–230 (2016)
6. A.M. Zaferiou, R.R. Wilcox, J.L. McNitt-Gray, Modification of impulse generation during pirouette turns with increased rotational demands. J. Appl. Biomech. **32**(5), 425–432 (2016)
7. A.M. Zaferiou, H. Flashner, R.R. Wilcox, J.L. McNitt-Gray, Lower extremity control during turns initiated with and without hip external rotation. J. Biomech. **52**, 130–139 (2017)
8. S. Sahrmann, *Diagnosis and Treatment of Movement Impairment Syndromes* (Mosby, St. Louis, 2002)
9. B.J. Erickson et al., Are the hamstrings from the drive leg or landing leg more active in baseball pitchers? An electromyographic study. J. Shoulder Elb. Surg. **26**(11), 2010–2016 (2017). https://doi.org/10.1016/j.jse.2017.06.041

Kineto-Dynamic Modeling of Human Upper Limb for Robotic Manipulators and Assistive Applications

Giuseppe Averta, Gemma C. Bettelani, Cosimo Della Santina, and Matteo Bianchi

Abstract The sensory-motor architecture of human upper limb and hand is characterized by a complex inter-relation of multiple elements, such as ligaments, muscles, and joints. Nonetheless, humans are able to generate coordinated and meaningful motor actions to interact—and eventually explore—the external environment. Such a complexity reduction is usually studied within the framework of synergistic control, whose focus has been mostly limited on human grasping and manipulation. Little attention has been devoted to the spatio-temporal characterization of human upper limb kinematic strategies and how the purposeful exploitation of the environmental constraints shapes human execution of manipulative actions. In this chapter, we report results on the evidence of a synergistic control of human upper limb and during manipulation with the environment. We propose functional analysis to characterize main spatio-temporal coordinated patterns of arm joints. Furthermore, we study how the environment influences human grasping synergies. The effect of cutaneous impairment is also evaluated. Applications to the design and control of robotic and assistive devices are finally discussed.

Keywords Motor control · Grasp · Hands · Daily living activities · Upper limb · Functional analysis · Human-inspired robotics

1 Introduction

Anaxagoras thought that men are the most intelligent of the animals, because they have hands. The main concept behind this sentence is that one of the most important activities of our body is related to the interaction with, and the manipulation of the environment and objects.

The whole upper limb has a marvelous structure, with hundreds of muscles and bones and sensors. This guarantees a large variety of possibilities, typically sum-

G. Averta (✉) · G. C. Bettelani · C. Della Santina · M. Bianchi
Centro di Ricerca Enrico Piaggio, Universita di Pisa, Pisa, Italy
e-mail: giuseppe.averta@ing.unipi.it

R. Vinjamuri (ed.), *Advances in Motor Neuroprostheses*,
https://doi.org/10.1007/978-3-030-38740-2_3

marized in four functional activities: (1) sensation, (2) holding, (3) manipulation, and (4) communication. Nonetheless, the human brain is able to control such a complexity and versatility in a very simple fashion [11]. The specific mechanisms underlying this simplification are still under investigation; however, it is nowadays broadly accepted that the overall movement behavior could be at least approximated through a reduced dimensionality control space [46, 52, 61, 69], typically named synergistic control space.

Indeed, most of the movements of our limbs are generated through a coordinated and harmonic enrollment of different Degrees of Freedom (DoFs) of the kinematic chain. A description of synergistic patterns has been reported at different level of the motion control natural architecture, i.e., neural [73], muscular [26, 72], kinematic [41, 50, 62], and kinetic [75]. For a review on these topics, the interested reader may refer to [65]. Focusing on the kinematic level, these observations pointed toward the idea that most of the limb postures variability can be explained by the combination of a reduced set of eigenvectors, i.e., covariation patterns of joints, also called principal components or postural synergies, while the remaining components of the synergistic basis are likely enrolled for the description of more complex tasks and fine movements, e.g., for haptic exploration [74] and contact forces distribution [6].

Several statistical methods have been used to describe kinematics synergies, e.g., principal components analysis (PCA), single value decomposition (SVD), functional PCA (fPCA), and non-negative matrix factorization (NNMF). In [62], PCA applied to a dataset of grasping poses revealed that the first three PCs explained $\sim$90% of the total hand poses variability, while the first two PCs accounted for $\sim$84%. These findings were then confirmed in other studies that take into account real object grasps [43] and the inter-digit coordination occurring during the whole grasping procedure.

Despite the huge attention devoted to the analysis of hand motor control, little has been done to investigate synergistic control of the whole upper limb. In this chapter, we aim at bridging this gap moving from hand synergies to upper limb kinematics investigation, with the goal of unveiling the principal synergistic actuation pattern that underpin motion generation in space and time. At the same time, the avenue of soft, adaptable yet robust artificial grippers that can deform in a human-like manner to mold around different items and fully exploit environment to multiply their degrees of freedom, has motivated our scientific interest in understanding the role of environmental constraints in synergistic control of human hands. This chapter is organized around these two topics, and their potential implications for the design, the planning, and the control of assistive devices, co-bots, and autonomous robots.

2 Experimental Setup for Data Acquisition

A thorough investigation of main upper limb modes and the synergistic behavior of the human hand in interaction with the environment necessarily needs to move from accurate kinematic recordings. To achieve this goal, we employed a commercial motion tracking system (i.e., *Phase Space*®). Ten 480 Hz stereo-cameras tracked online the 3D position of active markers rigidly fastened to upper limb and hand links. We used 20 markers for upper limb acquisitions and additional 20 markers for hand kinematics tracking, four for each finger. We suitably designed and printed in ABS (see Fig. 1a) rigid supports to accommodate markers. Kinematic data were synchronized with force/torque sensors and collected through a C++ custom routine.

The experiment execution was recorded through two additional cameras, with the ultimate goal of visually comparing real and reconstructed movement. Tasks were repeated once using the free hand and once with subjects wearing ThimbleSenses in all fingertips (see Fig. 1e), i.e., wearable sensors that provide complete single-digit force/torque measurements and contact point estimation [9]. Since Thimblesense are composed by an external rigid shell worn at the fingertip level, their usage also provides cutaneous impairment. The latter aspect is very interesting to be investigated since it could provide useful guidelines for sensing soft robotic hands, as discussed later. All the shells are connected to the fingertips as for classic thimbles. The reader is invited to refer at [9] for additional details on the impairment effects of the shells. Force information provided by these sensors was also acquired to be used in future investigations. Force/torque information was also measured on the object/environment side. To this purpose, a sensorized surface (600×400 mm), endowed with a force-torque ATI mini45E sensor mounted as in Fig. 1f, was used to keep track of the interaction of the subjects with the table.

3 Modeling

3.1 Kinematic Model of Human Upper Limb

The description of human kinematics is a very challenging task, e.g., see [44, 51], not only for the number and the axis directions of the Degrees of Freedom (DoFs), but also because the capabilities of motion measurement/reconstruction need to be carefully considered. For this reason, a good trade-off between accuracy and reasonable computational time is represented by a model with seven revolute joints and three rigid links [10]. In this chapter we will refer to joint angles $x = [q_1, \ldots, q_7]^T$ as:

- q_1 abduction-adduction of the shoulder;
- q_2 flexion-extension of the shoulder;
- q_3 external-internal rotation of the shoulder;
- q_4 flexion-extension of the elbow;
- q_5 pronation-supination of the elbow;
- q_6 abduction-adduction of the wrist;
- q_7 flexion-extension of the wrist.

Corresponding axes are depicted in Fig. 2.

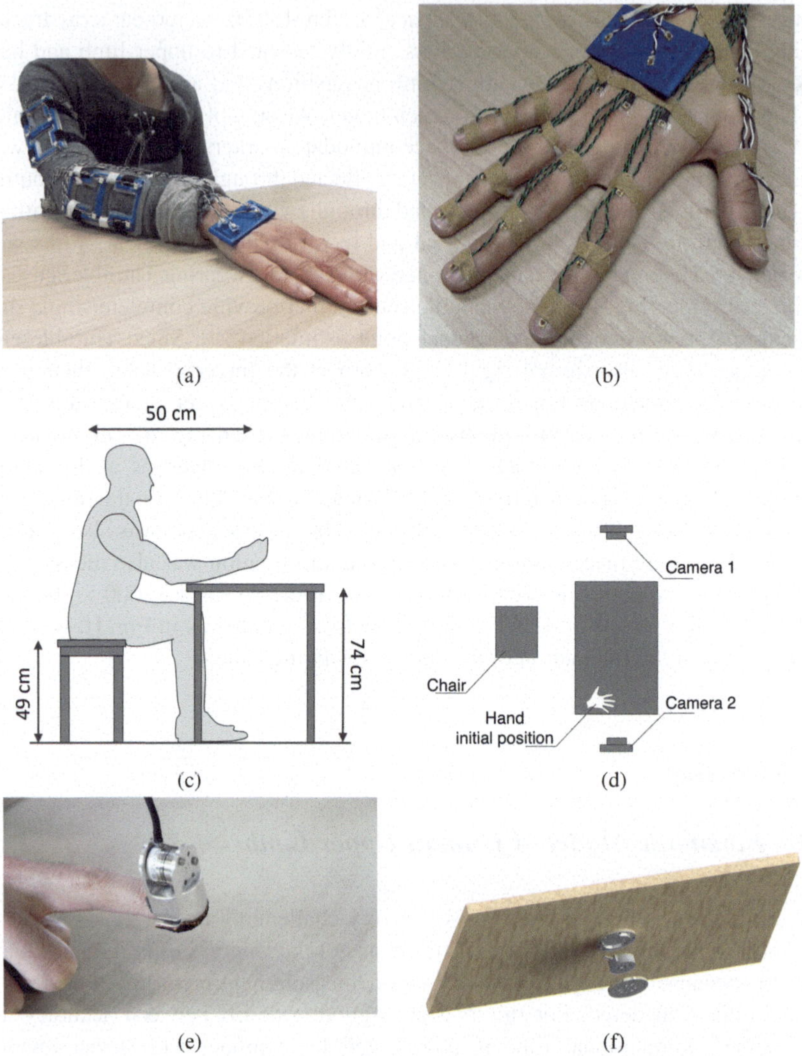

Fig. 1 In (**a**) and (**b**) we report the markers accommodation for upper limb and hand, respectively. In (**c**) and (**d**) we show a schematic representation of the experimental setup. Two cameras are included to record the scene. In (**e**) we show the shells at fingertip level, which produce tactile impairment and in (**f**) an exploded view of the sensorized table

As a consequence, the forward kinematic is completely described by five reference systems:

- S_{ref}, with the origin in O_{ref} and fixed to the epigastrium;
- S_S, with the origin in O_S, Center of Rotation (CoR) of shoulder joints and fixed to the arm;

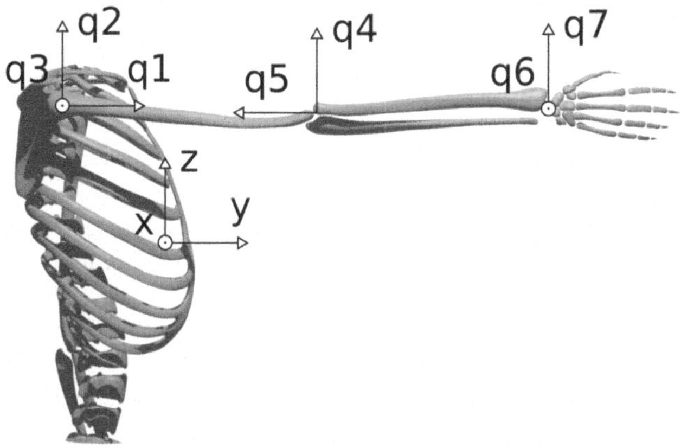

Fig. 2 The kinematic model used in this chapter to model the upper limb

- S_E, with the origin in O_E, CoR of elbow joints and fixed to the forearm;
- S_W, with the origin in O_W, CoR of wrist joints and fixed to the hand;
- S_H, with the origin in O_H and fixed to the hand.

The relative rigid transform between the reference systems are

- $T_{O_{ref}O_S}$ between S_{ref} and S_S;
- $T_{O_SO_E}$ between S_S and S_E;
- $T_{O_EO_W}$ between S_E and S_W;
- $T_{O_WO_H}$ between S_W and S_H.

We use here the *product of exponentials* (POE) formula [16] to parameterize the i-th segment

$$g_{O_{ref}O_j}(\theta) = \left[\prod_{k=1}^{j} e^{\hat{\xi}_k\theta_k}\right] g_{O_{ref}O_j}(0)$$

where $\hat{\xi}_k$ are the joints twists that define the kinematic chain, $\theta = [\theta_1, \ldots, \theta_k, \ldots, \theta_j]^T$ are the second kind exponential coordinates for a local representation of SE(3) (Special Euclidean group, 4×4 rototranslation matrices) for the j-th link, and $g_{O_{ref}O_j}(0)$ is the starting configuration. Additional details are reported in [39].

This model can be completely parametrized by 14 subject-specific coefficients, collected in a vector p_G: length of the arm and forearm links (2 pars); translation from the epigastrium to the CoR of the shoulder (3 pars); translation from the CoR of the shoulder to the center of arm marker support (3 pars, see the following subsection for details); translation from the CoR of the elbow to the center of forearm marker support (3 pars, see the following subsection for details); translation from the CoR of the wrist to the center of hand marker support (3 pars, see the following subsection for details).

3.1.1 Markers Placement

As already introduced previously, the motion of arm links is tracked through active optical markers, fastened to the upper limb through custom rigid supports. A redundant configuration of markers is used to avoid occlusions and improve tracking quality, i.e., six markers for arm, six markers for forearm, four markers for chest, and four for the hand (see Fig. 1a). The position of each single marker is known w.r.t. the center of the rigid support, while the relative configuration of the support w.r.t. the whole kinematic chain is one of the parameters to be calibrated for each subject.

3.2 Kinematic Model of the Human Hand

Regarding the hand, we used a kinematic model with 20 DoFs (see Fig. 3a). The four long fingers are moved by four independent joints:

- two DoFs for the metacarpo-phalangeal joint (for flexion-extension and abduction-adduction);
- one DoF for the proximal joint (flexion-extension);
- one DoF for the distal joint (flexion-extension).

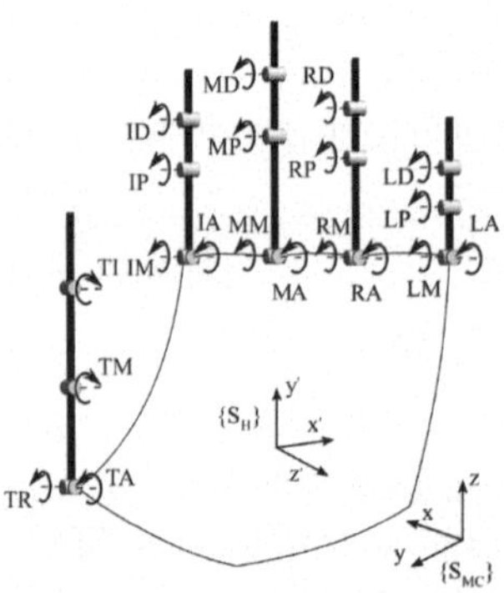

DoFs	Description
TA	Thumb Abduction
TR	Thumb Rotation
TM	Thumb Metacarpal
TI	Thumb Interphalangeal
IA	Index Abduction
IM	Index Metacarpal
IP	Index Proximal
ID	Index Distal
MM	Middle Metacarpal
MA	Middle Abduction
MP	Middle Proximal
MD	Middle Distal
RA	Ring Abduction
RM	Ring Metacarpal
RP	Ring Proximal
RD	Ring Distal
LA	Little Abduction
LM	Little Metacarpal
LP	Little Proximal
LD	Little Distal

(a)

Thumb (4 axis, 4R)

j	θ	d	a	α	offset
1	q_1	0	0	$-\frac{\pi}{2}$	0
2	q_2	0	a_2	$-\frac{\pi}{2}$	0
3	q_3	0	a_3	0	0
4	q_4	0	a_4	0	0

Long Finger (4 axis, 4R)

j	θ	d	a	α	offset
1	q_1	0	0	$-\frac{\pi}{2}$	0
2	q_2	0	a_2	0	0
3	q_3	0	a_3	0	0
4	q_4	0	a_4	0	0

(b)

Fig. 3 Model of hand kinematics. The 20 joints considered include the 15 DoFs of the hand model described in [62]. (**a**) Graphics of the hand kinematics; (**b**) Denavit–Hartenberg parametrization of the fingers

The thumb is moved by four independent joints:

- two DoFs for the trapeziometacarpal joint
- one DoF for the metacarpo-phalangeal joint;
- one DoF for the interphalangeal joint;

It is worth noticing that such a description shares all the 15 DoFs used in [62], thus enabling the comparison of the analysis results with classic description of postural grasp synergies. More details in the following of the chapter.

4 Motion Identification

The tracking of markers in the 3D space provides information on the Cartesian space. To infer from this the actual joint values it is necessary to solve a motion reconstruction problem. In this chapter we develop a procedure consisting of two phases: as a first step, since the length of the links is subject-specific, an optimization is carried out to calibrate the vector of parameters p_G. Then, the calibrated model is used within a frame-by-frame identification procedure performed through an extended Kalman filter (EKF).

4.1 Model Calibration

The optimization of the vector of parameters p_G is achieved by solving the following constrained problem:

$$(x^*, p_G^*) = arg \min_{x_k \in D_x, p_G \in D_p} \frac{1}{2} \sum_{k=1}^{N_p} r_k^T r_k,$$

where $r_k = r_k(x_k, p_G) := y_k - f(x_k, p_G)$ is a residual function, y_k is the vector of the measured markers position, x_k is an estimation of joint angles, D_x is the range of motion of the joints, D_p is a constraint on the maximum allowable variation of the kinematic parameters w.r.t. the initial manual measurement, and $f(x_k, p_G)$ is the estimated markers position via forward kinematics.

4.2 Motion Identification

The joint values in time are then estimated through an EKF. Let us consider the model as an uncertain noisy process, where x_k is the state at time frame k, y_k is the 3D position of markers, $f(x_k)$ is the forward kinematics, and w_k and v_k are the Gaussian noises of process and observation, with covariance Q_k and R_k, respectively.

This description can be formalized as in the following classical form:

$$\begin{cases} x_k = x_{k-1} + w_k \\ y_k = f(x_k) + v_k \end{cases} \tag{1}$$

The EKF implementation will result in the following procedure: given the state x_{k-1}, a prediction at the time frame k is calculated as $\hat{x}_{k|k-1} = \hat{x}_{k-1}$. Then, the prediction is updated by calculating $\hat{x}_{k|k} = \hat{x}_{k|k-1} + K_k \tilde{r}_k$. The amount of prediction correction of state estimation is calculated as product between the residual $\tilde{r}_k = y_k - f(\hat{x}_{k|k-1})$ and the Kalman gain K_k. The gain is calculated as product between $P_{k|k-1}$, the Jacobian matrix $H_k = \frac{\partial(f(x))}{\partial(x)}$, and the inverse of the residual covariance. The covariance matrices are tuned heuristically.

This two-phase procedure has been used for both the arm and the hand motion identification. However, given the lower distance between markers, the latter case is more prone to markers occlusion. To overcome this issue, the identification of fingers movement has been integrated with an additional constraint in the identification procedure. More specifically, we developed an adaptation procedure (online) of the observation noise covariance matrix with a scaling factor proportional to the number of consecutive missing values. In this way the observation noise covariance for the EKF is automatically increased in case of missing frames for the specific frame.

5 Principal Functions for Upper Limb Movement Generation

5.1 Experiments

It has been broadly discussed that, to correctly describe human movements in a general way, one of the key points is the number and variety of tasks considered in the dataset [50, 62, 71, 77]. In this work, we selected a set of 30 different daily living tasks, considering the main hand grasping poses (we referred to the grasping taxonomies discussed in [25, 36]) and ranging the whole upper limb workspace [1, 48, 54].

Driven by neuroscientific findings [24, 43], the 30 tasks can be organized in three classes: intransitive actions, i.e., gestures without interaction with objects; transitive actions, i.e., grasps of objects; tool-mediated actions, i.e., grasp of objects used as a tool to interact with another object. The complete list of actions is reported in [5].

Seven right-handed participants (5 M, 2 F, age ranging between 20 and 30 y.o.) were asked to randomly perform all the action of the protocol three times.

The quality of the estimated angular values—at each time frame *k*—was evaluated through the mean squared error (MSE) R_k as

$$R_k = \frac{1}{N_{markers}} ||(y_k - f(\hat{x}_{k|k}))||,$$

where $N_{markers}$ is the marker's number, y_k is the marker's position, and $f(\hat{x}_{k|k})$ is the estimation of markers position via forward kinematics and the joint angles estimated via EKF. In our experiments we had R_k typically $\approx$1 cm, with lower errors at the hand and forearm level ($\approx$0.5 cm) and larger errors at the arm level ($\approx$1.8 cm).

5.2 *Data Analysis*

To identify the functional principal components of motor behavior we leverage on the functional PCA, a statistical method that identifies the underpinning basis functions of a generic time-related dataset [58, 59]. Since the comparison between dataset elements is performed in time, a preliminary step required is represented by the time warping of the single entries. This is necessary to prevent—in the analysis—effects due to misalignment in time or different task velocity execution.

More specifically, time warping is a technique used to enable the comparison between different time-varying samples. Let us assume two time series v_1 and v_2; the affinity between v_1 and v_2 is maximized by solving the following optimization problem:

$$(S, T) = arg \min_{S>0,T} (||v_1(t) - v_2(St - T)||), \tag{2}$$

where S is a velocity scaling factor and T is a time-shifting. The general, dynamic, application of time warping accepts as a solution non-linear scaling function, i.e., $S(t)$ and $T(t)$. However, to preserve the signal shape, in our implementation we leverage on the solution of 2 with the additional constraint of S and T constant values. All the elements are warped w.r.t. a common reference, resulting in a new dataset with time-coherent elements.

5.2.1 A Functional Extension of PCA

As previously discussed, the main goal of this section is the description of the upper limb movements under a functional point of view. One solution to this problem is represented by functional PCA (fPCA).

The main idea behind fPCA is conceptually similar to the classic PCA analysis and can be formalized as in the following.

Given a dataset of functions $X = [x_1, \ldots, x_i, \ldots, x_N]$, it is possible to calculate a first functional principal component $\xi_1(t)$ as the function that maximizes the following objective function:

$$\sum_{i=1}^{N} f_{i1}^2, \quad \text{where} \quad f_{i1} = \int \xi_1(t) x_i(t) dt, \tag{3}$$

$$s.t. \int \xi_1^2(t) dt = ||\xi_1|| = 1. \tag{4}$$

The second functional principal component $\xi_2(t)$ maximizes the following objective function:

$$\sum_{i=1}^{N} f_{i2}^2, \quad \text{where} \quad f_{i2} = \int \xi_2(t) x_i(t) dt, \tag{5}$$

$$s.t. \quad ||\xi_2|| = 1 \quad \text{and} \quad \int \xi_2(t) \xi_1(t) dt = 0. \tag{6}$$

Higher order PCs can be defined recursively as for Eq. 5. The practical implementation, which bypasses the optimization, is listed in the following:

1. Let us assume a dataset of functions $X = [x_1, \ldots, x_i, \ldots, x_N]$, where $x_i \in \mathbb{R}^{N_{DoF}, T}$, T is the number of time frames, and N_{DoF} is the number of Degrees of Freedom. For each joint, evaluate the mean $\bar{x} = [\bar{x}_1 \ldots \bar{x}_j \ldots \bar{x}_{N_{DoF}}]'$ as $\bar{x_j} = \frac{1}{N} \sum_{i=1}^{N} x_{ij}$;
2. Remove the mean from each element of the dataset $\tilde{x_i} = x_i - \bar{x}$;
3. Define a basis of functions, large enough to consider all possible modes of variations. Typical definitions are exponential functions, splines, Fourier basis [57–59];
4. Given the basis elements $b_1, \ldots, b_N$, calculate each element of the dataset as a combination of basis elements $\tilde{x}_i = \sum_{k=1}^{N} \theta_k b_k$;
5. Each element of the dataset can be encoded through a vector of coefficients $\Theta = (\theta_1, \ldots, \theta_N)'$;
6. Calculate the principal components ξ_i of the new dataset of coefficients;
7. Each PC will results into a corresponding functional PC, i.e., $x_{rec} = \bar{x} + c_1\xi_1 + c_2\xi_2 + c_3\xi_3 + \ldots$;
8. The variance explained by each fPC is calculated as the normalized eigenvalue of the covariation matrix.

5.3 *Results*

We evaluated the functional principal components on the dataset introduced in the previous sections. The basis is defined with 15 fifth order spline [37]. More specifically, each basis element is defined as piecewise polynomial functions. Each piece of the function is of the form:

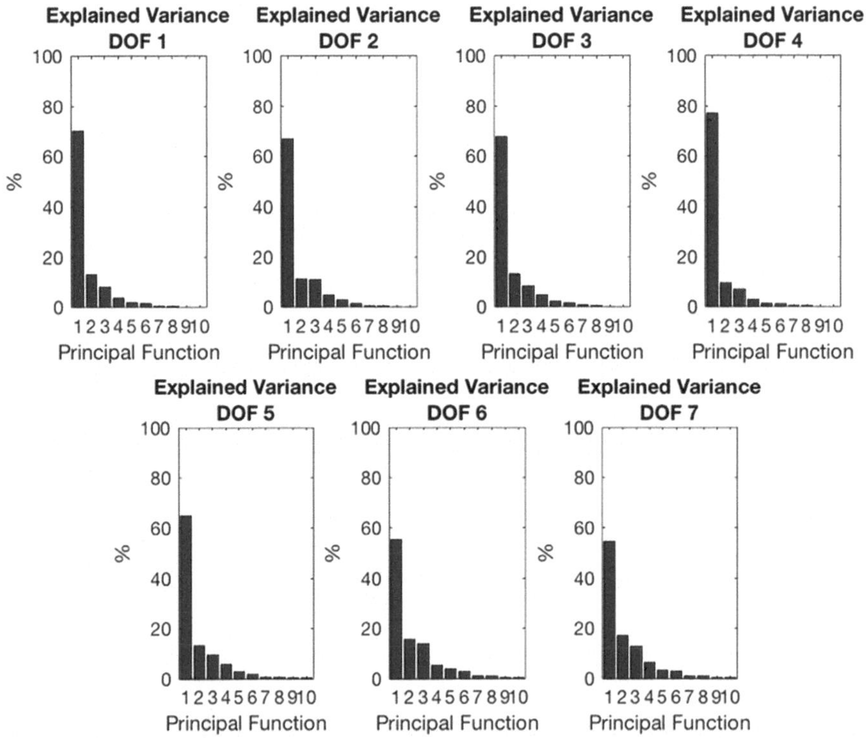

Fig. 4 Variance explained—for each joint—by fPCs

$$s_k(t) = \sum_{i=1}^{5} a_{ik}(t - t_k)^i, \tag{7}$$

where t_k is the kth knot.

Our results revealed that the first fPC accounts for 60–70% of the total joint dataset variation (see Fig. 4). The variability explained by the first component is higher for the DoFs related to the shoulder and elbow movements, while is lower for forearm pronation-supination and for the wrist DoFs. It is also worth noticing that a very reduced number of fPCs is typically sufficient to approximate most of the tasks considered in this study. Indeed, the first three fPCs explain more than 84% of the total variability for all the DoFs.

Figure 5 reports the first three fPCs as a modulation of the mean function. We invite the reader to notice that the cardinality of the specific fPC is related to the complexity of the resulting modulation. This points toward the idea that increasing the number of enrolled fPCs can result in higher overall complexity of the movement.

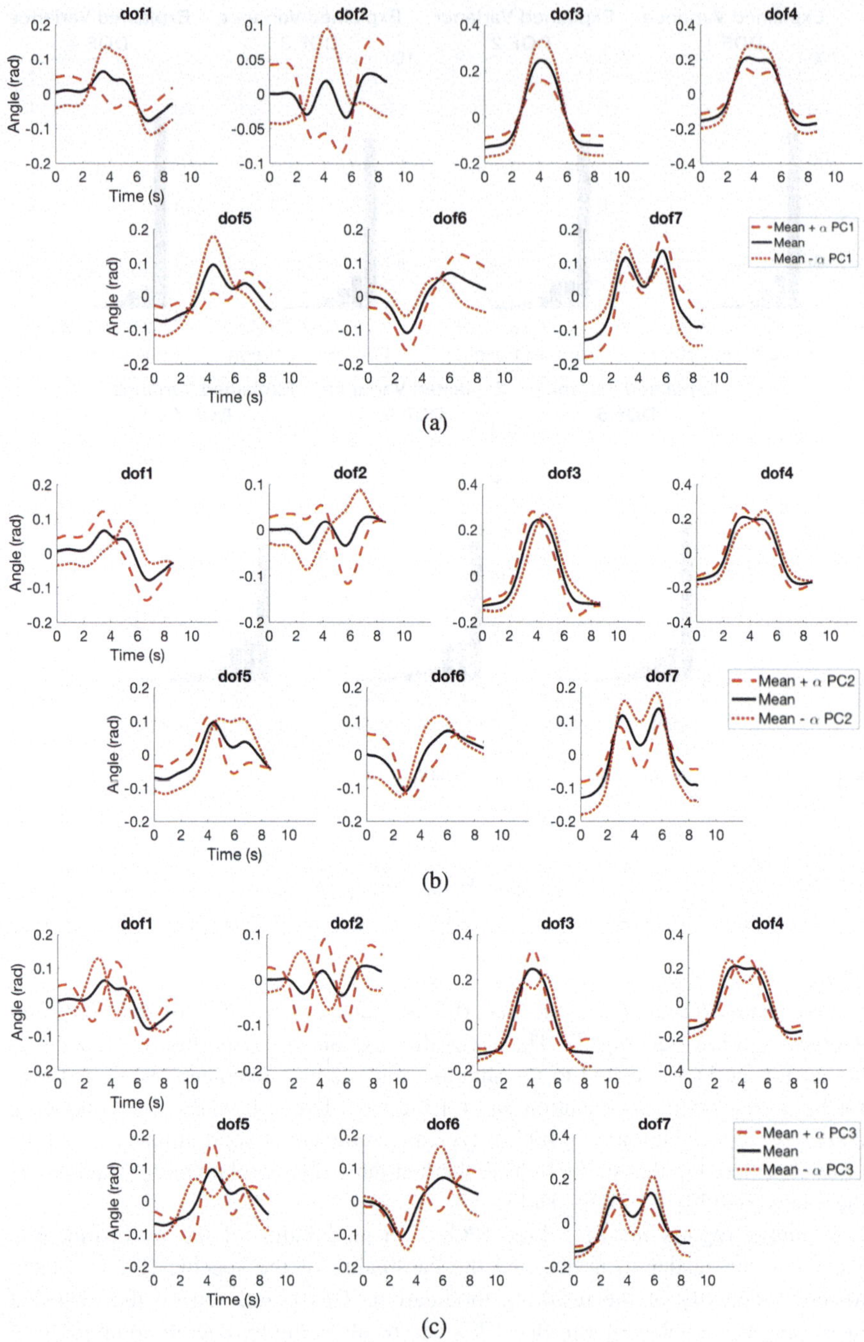

Fig. 5 First three fPCs reported in red as modulation w.r.t. the mean function (in black). (**a**) First fPC, (**b**) second fPC, (**c**) third fPC

The performances of signal reconstruction are reported in Fig. 6a, where a sample element of the dataset is approximated through one, two, and three fPCs. An index of reconstruction error is evaluated as:

$$E_{RMS} = \sqrt{\frac{1}{N_{DoF}} \sum_{i=1}^{N_{DoF}} \left[\sqrt{\frac{1}{N_{frames}} \sum_{j=1}^{N_{frames}} (x - x_{rec})^2} \right]^2}, \tag{8}$$

where x is the original element and x_{rec} is the fPC-based approximation.

Figure 6b reports the normalized reconstruction error, evaluated as $E_{RMS}/\max(E_{RMS})$, while increasing the number of fPCs enrolled. The initial value refers to the mean function only ($E_{RMS} = 0.6$ rad). With only one fPC the E_{RMS} value is lower than 0.2 rad. It is also worth noticing that the overall reconstructed movement preserves an high level of anthropomorphism and realism.

6 Postural Hand Synergies During Environmental Constraint Exploitation

This section deals with the investigation of a synergistic behavior that explains the hand postures during activities of environmental constraint exploitation (ECE) [28], with and without the effect of cutaneous impairment.

To address this goal, we carried out experiments with six healthy subjects (3 F, 3 M, age ranging between 23 and 27 y.o.), who were asked to reach and grasp set of objects from a table. The tasks considered were selected as a trade-off between protocol complexity and richness of tracked kinematic behavior [35]. For each task, repeated twice and randomly selected, subjects were instructed to reach the object placed on the sensorized table, grasp it, lift (∼20 cm height), hold (∼1 s), put it back on the original position, and return to the starting position.

The acquisitions were performed in two different experimental conditions: with and without tactile impairment. More specifically, in the first case subjects were asked to wear rigid shells at the fingertips level to prevent tactile feedback. Hand postures were processed via principal component analysis (PCA) [64] to identify the synergistic behavior in hand posture codification. The analysis was performed on pre-shaping postures (i.e., before the contact with the object) and during the actual contact/interaction with the environment. Signals collected during the experiments are hand postures through 24 active markers position, force/torque of the interaction with the table (F/T ATImini45).

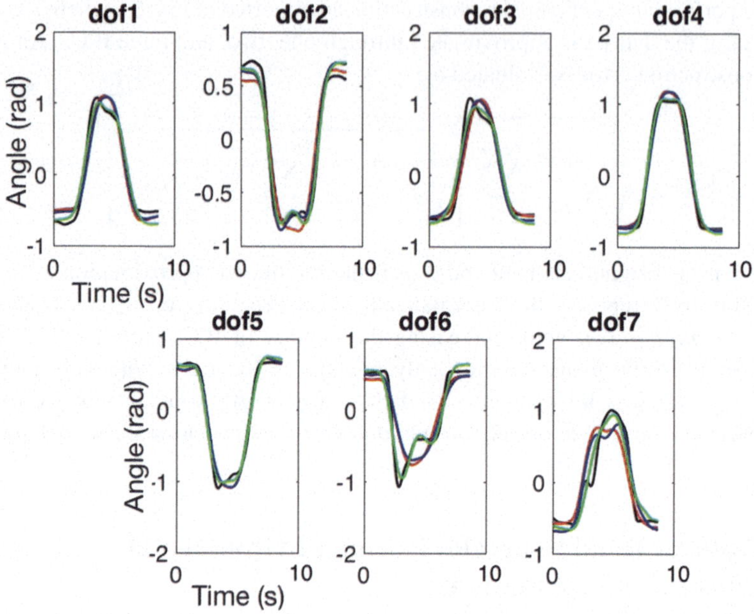

(a) Trajectories reconstruction

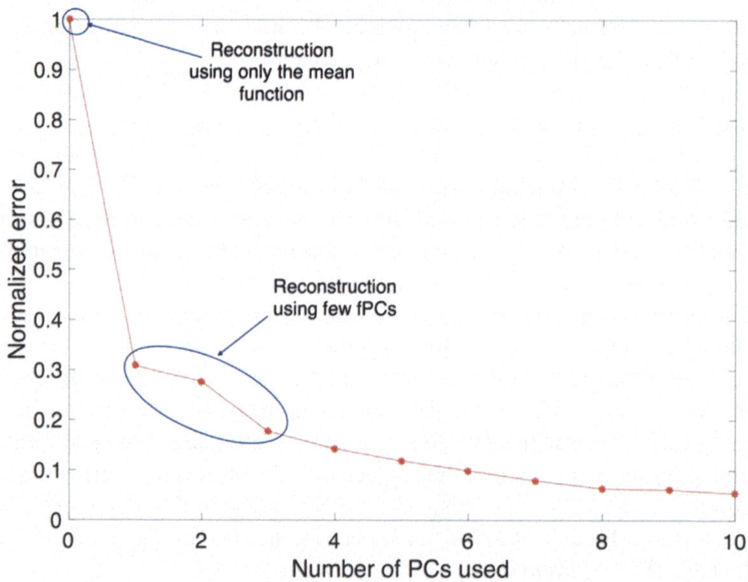

(b) Normalized Reconstruction Error (RMS)

Fig. 6 (**a**) In black real acquisition; in red the reconstruction using one fPC; in blue the reconstruction using two fPCs; in green the reconstruction using three fPCs. (**b**) Reconstruction error vs. number of fPCs considered to approximate the signal

6.1 Pre-processing

Data of force/torque from the sensorized table are filtered via moving average through the Savitzky–Golay method [66]. The filter window width is heuristically tuned as the 1.5% of the total signal length. Then, leveraging on the knowledge of the surface geometry, it is possible to evaluate the centroid of contact, which will be used in the following of this section for contact triggering.

Indeed, postural data are segmented in three phases (see also Fig. 7 for a visual description):

1. pre-shaping, the phase in which the hand has reached the object and is shaped as to interact with the object/environment;
2. contact, the phase in which the constraint is exploited to grasp/manipulate the object;
3. post-contact, the phase in which the object is grasped and lifted from the table.

These phases are identified by looking at the signals recorded by the sensorized surface. The transition from first to second phase is triggered by the contact detection between hand and table/object (i.e., looking to the increasing force and force derivative). The transition from second to third phase is identified as the first time frame in which the force returns to zero.

As previously stated, the main objective of this section is to identify a subspace of reduced dimensionality for hand postures. To achieve this goal, principal component analysis (PCA) has been proven as an effective tool [7, 50, 62, 64, 70]. More specifically, given a dataset of postures described by a matrix of correlation C and a mean m, PCA identifies an orthonormal basis of data space, which first element S_1 is the direction that explains the largest percentage of data variability. Each successive principal component S_i explains the highest variability under orthogonality constraint.

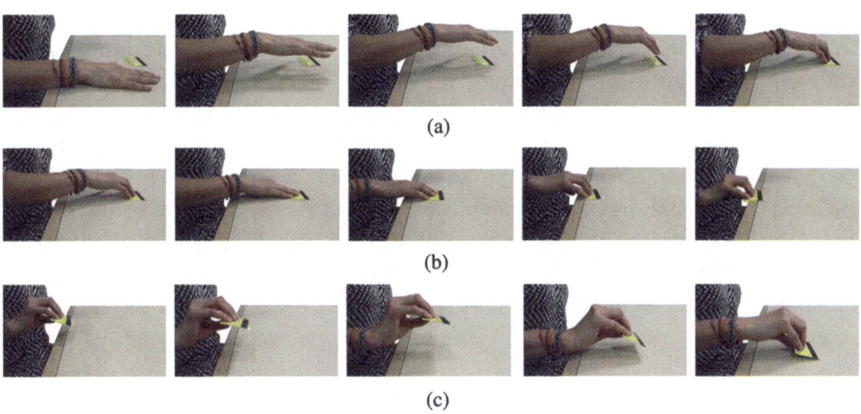

Fig. 7 Sample of environment exploitation to enhance grasping and manipulation capabilities. (**a**) is the *pre-shaping* phase, (**b**) is the *during contact* phase, and (**c**) is the *post-contact* phase

PCA is typically implemented as the singular value decomposition of the correlation matrix, i.e., calculating the orthonormal matrix Σ that brings C in Jordan form via the similitude $\Sigma^T C \Sigma$. In this implementation, the principal components are the column of $\Sigma = [S_1, \ldots, S_n]$, and the explained variance is the normalization of the corresponding eigenvalues [45].

It is worth noticing that the comparison between experimental conditions is enabled by the usage of calibrated kinematic model (see previous sections of this chapter). To quantify the similarity, we then quantified the normalized absolute value of the dot product between vectors. Similarities have been quantified between different experimental conditions and with the results associated with classical grasping analysis [62]. As discussed previously, the kinematic model employed in [62] implement a reduced number of DoFs (15, see also Sect. 3.2). In this case, to compare vectors we used the projection of the full hand configurations in the corresponding low-dimensional space (i.e., from $\mathbb{R}^{20}$ to $\mathbb{R}^{15}$). This resulted in simply neglecting values corresponding to ID, MD, RD, and LD in Fig. 3a (see also [39]).

6.1.1 Pre-shaping Analysis

The first phase of the motion that is worth analyzing is the pre-shaping. Indeed, in [62] the authors took out the effect of physical interaction by asking subjects to grasp imagined objects. Here, we aim at achieving the same goal by focusing the analysis on the last-hand pose before the contact with the environment, i.e., the last pose in pre-shaping phases when a purposeful interaction with the environment is planned. We separately analyzed the two experimental conditions of impaired and unimpaired tactile feedback. The two datasets are composed of 252 poses.

6.1.2 Contact Analysis

To evaluate whether kinematic covariation patterns are also noticeable during actual exploitation of the surface we performed PCA also on the data collected during the contact phase—considering separately the two experimental conditions of impaired and unimpaired tactile feedback. To quantify the effect of cutaneous impairment, we also calculated the mean amount of time in which subjects remained in contact with the table, mean time for task accomplishment, and the mean norm of interaction forces, by averaging the corresponding values for every subject and every object.

The considered datasets are composed of a variable number of poses, due to the variability in time execution of the task (we recorded 40 postures per second of task execution).

6.1.3 Differences Between Pre and During Contact

We also considered the persistence of the same covariation patterns of hand posture before and during the contact with the environment for both free and impaired hands. This was performed through the index based on the dot product between synergies, as previously discussed.

6.2 Results

6.2.1 Pre-shaping Analysis

Figure 8 reports the variance explained by the PCs for the analysis regarding the pre-shaping phase. For the unimpaired case, the first synergy explains about the 54% of the total dataset variability, while the first three PCs account for more than 72%. Figure 9 shows a graphical representation of the first three resulting postural

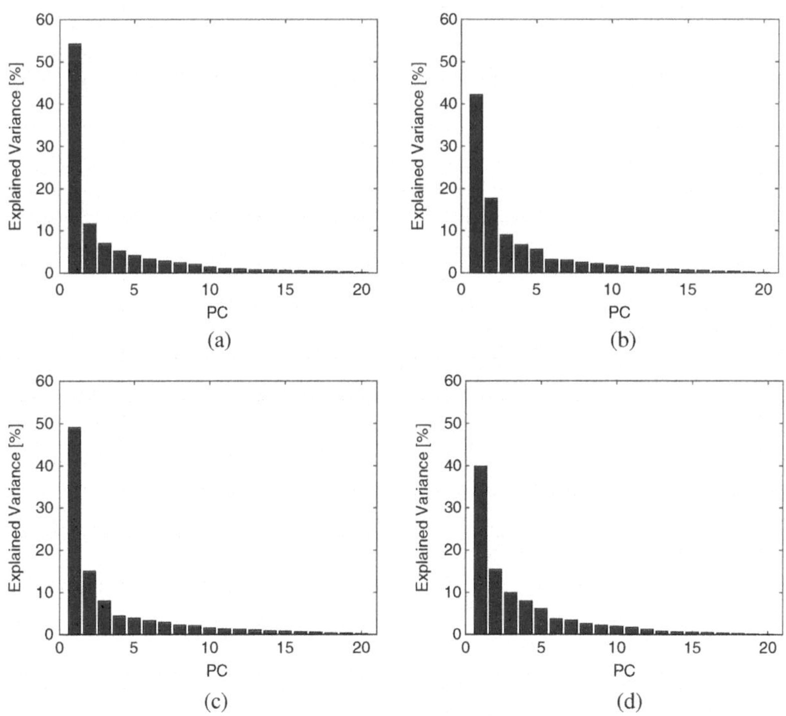

Fig. 8 Variance explained by the principal components during the preshaping and contact phases, with and without tactile impairment. A marked synergistic behavior is noticeable in all the cases. (**a**) Unimpaired, pre-shape. (**b**) Impaired, pre-shape. (**c**) Unimpaired, contact. (**d**) Impaired, contact

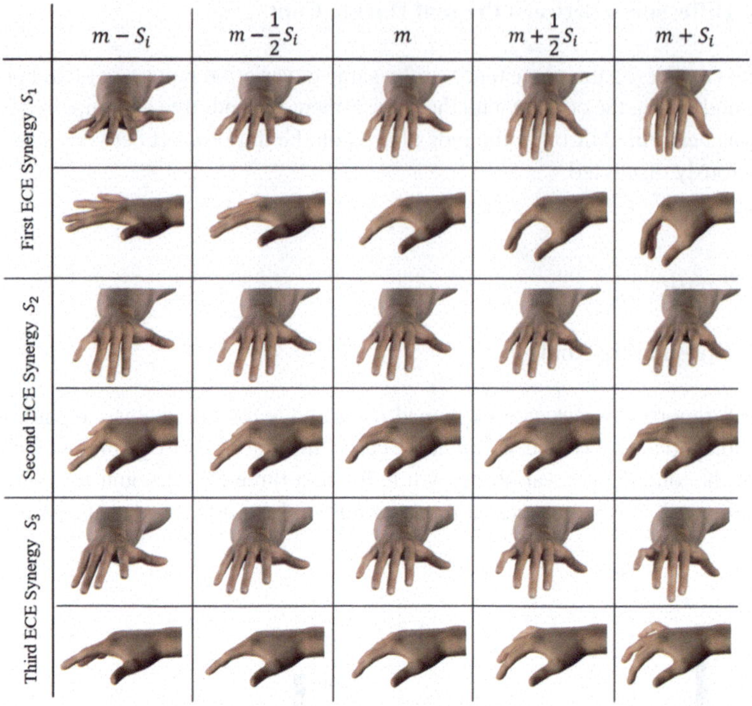

Fig. 9 Graphical representation of the hand configuration w.r.t. the mean pose, associated to the first three synergies during preshaping in unimpaired condition. Each column presents a different stage of the synergistic posture generation, obtained by summing the hand mean configuration m, to the synergy vector S_i of the i-th synergy

synergies. Regarding the impaired case, we observed that the first synergy accounts for the 42% of the total variability, while three principal components account for more than 68%. It is worth noticing that, for our dataset, the first synergies account for higher percentages of variance in the unimpaired w.r.t. the impaired condition. Table 1 lists the numerical values of the first ECE synergy of pre-shaping, with and without tactile impairment, in comparison with the first grasp synergy reported in [62]. Figure 10 presents the movement corresponding to the second synergy of pre-shaping with tactile impairment.

In all the considered experimental conditions, the first principal component encodes an opening–closing pattern for the whole hand, while the second principal component corresponds to a flexion of the distal joints—mostly of index and medial fingers—and a flexion of the thumb. The third synergy is similar to the second, with a prominent effect on the little and ring fingers.

Figure 11a shows the dot product between the first grasp synergy in [62] and the ones found in this work for the two experimental conditions. It is worth noticing the strong consistency between the first synergy in the different conditions (≥0.9), which such similarity is reduced for the second synergy, as shown in Fig. 11b, and

Table 1 Numerical values of the first synergy of Grasp [62] and of Environmental Constraint Exploitation, with and without impairment, before and after contact

DoFs	Grasp	Unimpaired pre-shape	Impaired pre-shape	Unimpaired contact	Impaired contact
TA	−0.43	−0.14	−0.15	−0.12	−0.15
TR	0.29	0.31	0.35	0.30	0.34
TM	0.14	0.14	0.17	0.17	0.16
TI	0.03	0.04	0.05	0.05	0.09
IA	−0.13	−0.08	−0.12	−0.08	−0.11
IM	0.33	0.39	0.35	0.40	0.34
IP	0.15	0.16	0.16	0.17	0.20
ID	x	0.01	0.03	0.03	0.05
MA	x	−0.03	−0.08	−0.02	−0.06
MM	0.33	0.38	0.33	0.38	0.32
MP	0.16	0.27	0.27	0.27	0.30
MD	x	0.04	0.06	0.05	0.08
RA	0.06	0.00	−0.02	0.02	−0.02
RM	0.40	0.44	0.37	0.43	0.32
RP	0.20	0.22	0.27	0.22	0.35
RD	x	0.04	0.06	0.05	0.11
LA	0.14	0.05	0.1	0.08	0.09
LM	0.37	0.43	0.42	0.41	0.37
LP	0.27	0.12	0.21	0.17	0.24
LD	x	0.02	0.04	0.02	0.08

We indicate with ‘x’ the DoFs that were not considered in [62]

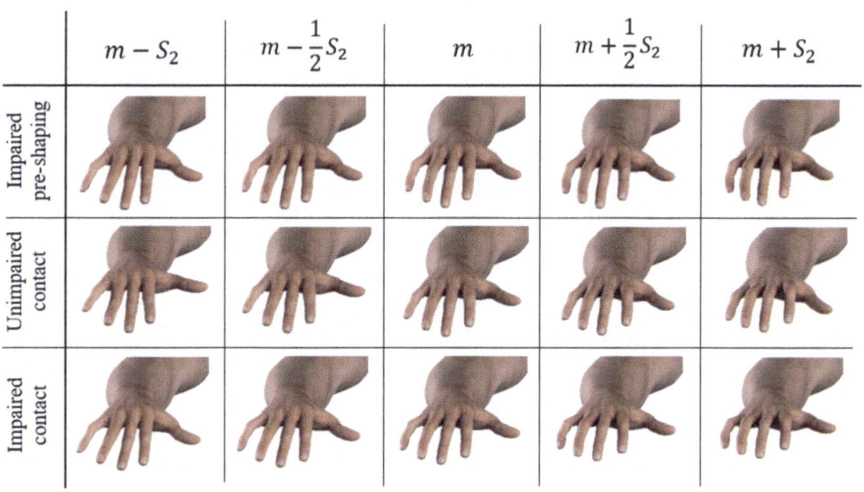

Fig. 10 Graphical representation of the hand shapes associated to the second ECE synergy. Pre-shaping impaired, contact unimpaired and contact impaired conditions are considered. The mean posture is referred as m, the second synergy as S_2. We do not report here the first synergy for each condition, since there are not significant and visible discrepancies. The figure also shows a good coherence in the behavior described by the second synergy, among the considered conditions

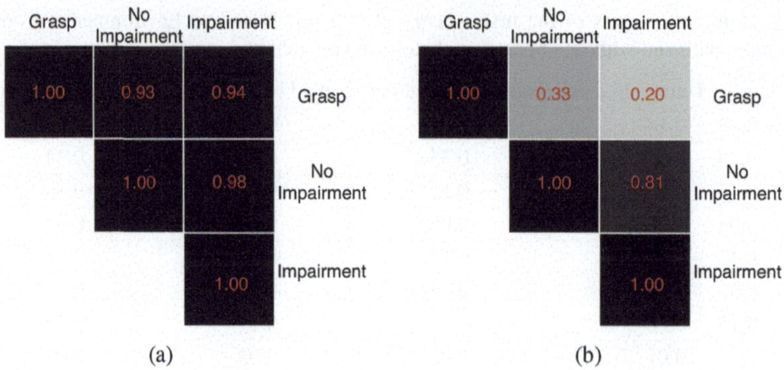

Fig. 11 Index of similarity between synergies, calculated as dot product. Values are codified in gray-scale: black is 1, i.e., very similar, white is 0, i.e., very different. ECE principal components in both the experimental conditions show high similarity, which, however, drops for higher order synergies. (**a**) First synergies. (**b**) Second synergies

so on for the other orders. Figure 11 shows also a high correlation between pre-shaping synergies in the two experimental conditions. Indeed, it can be observed that the introduction of tactile impairment does not alter the first two synergies during the pre-shaping phase. The similarity strongly drops for higher order synergies, reaching 0.36 for the third and 0.003 for the forth.

6.2.2 Contact Analysis

The analysis reported that, for the no-impairment condition, the first synergy accounts for about the 49% of the total dataset variability, while the first three explains more than 73%.

As already observed for the pre-shape case, these values are lower for the impaired case. Indeed, in this case we had a that the 39% of the dataset variance can be explained by the first synergy alone, while the first three account for the 65% of the total variability. Table 1 reports the numerical values of first ECE synergy in the contact phase, in the two experimental conditions, in comparison with the first grasp synergy [62]. Figure 10 shows the posture corresponding to the second synergy with and without tactile impairment.

Our study also reported that subjects, in case of tactile impairment, are in contact with the table for an average time of 4.2 ± 3.1 s, while in case of bare hand the average time is 2.4 ± 2.4 s. The task is completed in 13.9 ± 2.7 s in case of tactile impairment, while it is performed in 11.5 ± 2.0 s with the bare hand. Finally, also the contact force is different for the experiments considered, with a mean value for the impaired case of 23.2 ± 8.6 N and of 12.3 ± 5.7 N for the unimpaired case.

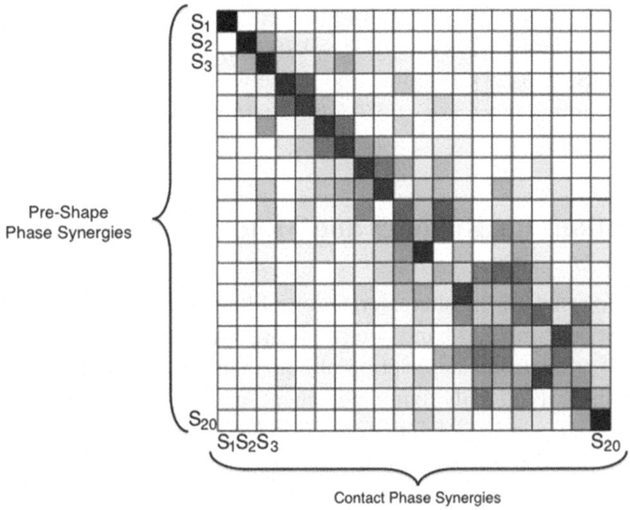

Fig. 12 Index of similarity—calculated via dot product—between the synergies in pre-shaping and in contact phases, considering the unimpaired condition. Values are codified in gray-scale: black is 1, i.e., very similar, white is 0. A tendency to maintain the first main components before and after the contact results clearly from this analysis. Results for the impaired case are analogous

6.2.3 Differences Between Pre and During Contact

Our results show a high similarity between the first principal components in the two experimental conditions w.r.t. the corresponding ones of the pre-shaping analysis. This similarity tends to decrease for higher order synergies, even if correlation values are ≥ 0.75 till ninth synergy. Values of the dot product are reported in Fig. 12 for the unimpaired case. These observations suggest that environmental constraints induce changes only for the high order synergies, leaving unaltered the main ones, regardless of availability of tactile input. Note that the first synergy is still strongly similar to the first grasp synergy [62].

6.2.4 Inference and Statistical Relevance

Results presented in this section rely on acquisition recorded on a cohort of six subjects. Despite the moderate number of participants, findings and analyses are in line with the existing literature in the field, see, e.g., [50, 53, 62]. To generalize, we performed additional statistical analyses and calculated the t-Student based confidence intervals (CI) with 95% probability. CI refers to dot products calculated on the synergies associated to different experimental conditions, and the ones associated to grasping [62]. More specifically, we had that the CI for the dot product between the first synergy for impaired and unimpaired conditions is $[0.81, 0.96]$; CI for the dot product between the first grasping synergy and the unimpaired first ECE synergy is $[0.88, 0.94]$; CI for the dot product between the first grasping synergy and the impaired first ECE synergy is $[0.79, 0.9]$. Regarding the analysis in the contact phase (Sect. 6.2.3), the dot product between the first synergy before and during contact results in a CI of $[0.97, 0.99]$ (unimpaired case).

7 Implications for Robotics and Motor Control

This chapter focused on the characterization of human movements, with particular attention for upper limb and hand control. More specifically, the first section faced the problem of a functional description of human arm movements, reporting on the functional complexity of daily living movements and on how it can be approximated through a reduced number of functional principal components. To achieve these results, we proposed the design of an experimental setup that includes kinematic recordings and other sensing modalities. Postural acquisitions rely on a 7 DoFs kinematic model and are quantified via a two-phase procedure for model calibration and angles identification.

We do believe that these findings could be used to provide a more accurate characterization of human upper limb functional principal modes, with potential applications in rehabilitations for the automatic recognition of pathological conditions [68]

At the same time, these results—and future developments—could be also useful to inspire the design, planning, and control of robotic manipulators. Indeed, leveraging on the fact that a very reduced number of functional principal components is sufficient to approximate with high precision very complex motions, it could be possible to envision the design of human-like movements trajectories exploiting on such functional basis. These strategies could be included in the combination of feedback and feedforward components, with direct application for the control of soft manipulators, i.e., robots able to implement natural and safe behaviors thanks to the introduction of compliant structures in the design which enable to achieve different—potentially variable—impedance characteristics. In these cases, standard feedback control has been proven to dramatically affect the compliant behavior of the system [3, 29]. To overcome these issues, in [29] a control strategy combining feedforward actions and low-gain feedback was proposed, with the ultimate goal of generating a human-like behavior via iterative learning [3]. In these applications, results presented in this chapter could be used to drive the design of the feedforward component.

Moreover, the usage of human-like primitives for robot control could represent an improvement of safety and effectiveness of Human–Robot Interaction (HRI). Indeed, it has been proved by different studies that anthropomorphism is one of the key enabling factors for a predictable, safe, acceptable, and successful HRI in many fields, e.g., for co-working applications and rehabilitative/assistive robotics [8, 32, 33, 60].

Furthermore, this experimental and analytical framework could be used to identify under-actuation strategies for human-inspired robotic devices. As an example, in [18], we employed this framework to quantify the role of wrist joints in the most common poses for grasping tasks. To do this, we calculated the principal components of the wrist pre-grasp poses and we observed that the flexo-extension DoF plays a dominant role. We used these results to inspire the design of an under-actuated robotic wrist, which is also compliant and enables to implement different schemes of under-actuation.

Future development of this work will target the usage of functional analysis as an enabling factor for a dynamic implementation of human upper limb modes for robotic applications. In addition, the integration of different sensing modalities, e.g., electro-encephalographic recordings, can be used to analyze neural correlates of motor control, thus providing useful insights for the development of Brain–Machine Interfaces [20, 21].

Regarding the characterization of hand postural synergies discussed in this chapter, we used a PCA-based analysis to describe the underlying synergistic description of hand motor control in case of environmental constraint exploitation, with particular focus on the pre-shaping and contact phases, and considering the effects of haptic feedback impairment. More specifically, we observed that the variability of hand poses (depicted in Fig. 8) suggests the effective presence of a reduced number of joint covariation patterns that account for a large part of the total variance. Indeed, one principal component explains up to 54% of the total dataset variability (for the unimpaired pre-shaping case), while the first three principal components account for more than 65% in all the tested conditions. We also observed that—in the case of contact with the environment, the variability accounted by higher order synergies increases. We believe that this is a direct effect of the physical interaction, which is responsible to a—not directly controlled—hand shaping. This is also in agreement with the observations reported in [63], where the analysis introduced in [62] was extended to real objects. Despite the fact that our inference analysis, discussed in Sect. 6.2.4, is limited to the first Principal Component, a series of characteristics of our dataset are worth of consideration, in accordance with existing neuroscientific results. Future work will target additional tasks and different experimental procedures to further investigate these aspects. Considering the data we collected, we observed high similarities between the first two synergies in the impaired and unimpaired conditions (see Figs. 10 and 11). This suggests that the effect of haptic impairment does not substantially alter the most important basis components of hand grasping, even if changes in the approaching strategies are still detectable (see [4]). It is worth discussing that subjects were aware of tactile impairment presence, then it is reasonable to expect that part of these differences is related to changes in motor planning as a consequence of feedback impairment. However, the dimensions of the collected dataset does not allow a strong statistical characterization of this behavior yet, and, to assess whether higher order components are primarily noise or rather they actively contribute to hand postures definition, in our future work we will resort to the usage of discriminant analysis and information theory, as done in [62]. Analogously, Fig. 12 reports a strong similarity in the first principal components related to pre-shaping and contact phases, respectively, while differences are noticeable for higher order components. The theory of the uncontrolled manifold [47, 67] hypothesizes that the central nervous system selects—within the space of joint angles—a reduced set of variables of interest and devotes the control to these, while purposefully leaving free the remaining Degrees of Freedom.

The persistence of the main principal components across the pre-shaping and contact phases can be reviewed following this theory by considering the first ECE synergies as the variables of interest for the specific task, which are preserved when external disturbances occur. The DoFs encoded by higher order synergies, instead, are left free to adapt to the external environment. It is also possible that this behavior is related to physical constraints embedded in the musculoskeletal system, as discussed, e.g., in [64]. Results depicted in Fig. 11 and Table 1 suggest that a strong similarity is evident between the first synergy related to the ECE strategies and the first grasp synergy presented in [62]. This may suggests the presence of underlying—general—synergies, integrated with task specific ones. This was initially proposed, e.g., in [40], and it is in agreement with experimental observations discussed in [70], where task independent synergies are estimated by a set of unconstrained tasks (see also CI estimates in Sect. 6.2.4). It is nowadays broadly accepted in literature that the results of neuroscientific analysis of human hand synergistic behavior can be successfully exploited and applied in robotics, to inform the design, planning, and control of mechatronic devices, with particular attention on grasping [12, 14]. One of the first attempts to exploit kinematic synergies in robotics is reported in [17], where the authors proposed to use grasp synergies as actuation patterns for an under-actuated hand. In [38] and [6] the role of postural synergies for the distribution of grasping forces is discussed. In [23] and later in [2, 49, 76], a low-dimensional space based on synergies is used to purposefully obtain effective pre-grasp configurations for a fully actuated robotic hand. More recently, synergy-inspired actuation framework have been combined with the introduction of physical compliance in the design [19, 22, 27, 30, 78] (see also [56] for a review), according to the soft-synergy paradigm introduced in [14]. The availability of elasticity in robotic hands also led to a shift under the planning and control point of view, see, e.g., in [13, 15, 31, 34]. The most direct implication of the results presented in this chapter could rely on the strong similarity between the first ECE synergy and the first synergy of grasping and the implementation of this joint covariation pattern as a degree of actuation can target the twofold goal of designing under-actuated robotic hands able to effectively grasp objects with and without the environmental constraints exploitation. To enhance hand capabilities beyond the first degree of actuation, we could include additional ECE synergies, possibly in combination with the grasp synergies (see [27, 42, 55]).

Referring to the differences introduced by tactile impairment, the principal kinematic patterns appear to remain preserved—at least for the gross movements. However, differences are observed in terms of time for task accomplishment and exchanged forces. This points toward possible sensing strategies for soft robotic hands, i.e., to effectively detect the contact with the object or the external environment, e.g., using IMU sensors, which may inspire the development of planning and control strategies aiming at minimizing exerted force on external objects.

Acknowledgements This research has received funding from the European Union's Horizon 2020 Research and Innovation Programme under Grant Agreement No. 688857 (SoftPro).

References

1. K. Abdel-Malek, J. Yang, R. Brand, E. Tanbour, Towards understanding the workspace of human limbs. Ergonomics **47**(13), 1386–1405 (2004)
2. H.B. Amor, O. Kroemer, U. Hillenbrand, G. Neumann, J. Peters, Generalization of human grasping for multi-fingered robot hands, in *2012 IEEE/RSJ International Conference on Intelligent Robots and Systems (IROS)* (IEEE, 2012), pp. 2043–2050. https://doi.org/10.1109/iros.2012.6386072
3. F. Angelini, C. Della Santina, M. Garabini, M. Bianchi, G.M. Gasparri, G. Grioli, M.G. Catalano, A. Bicchi, Decentralized trajectory tracking control for soft robots interacting with the environment. IEEE Trans. Robot. **34**(4), 924–935 (2018)
4. G. Averta, C. Della Santina, E. Battaglia, S. Ciotti, V. Arapi, S. Fani, M. Bianchi, From humans to robots: the role of cutaneous impairment in human environmental constraint exploitation to inform the design of robotic hands, in *2017 9th International Congress on Ultra Modern Telecommunications and Control Systems and Workshops (ICUMT)* (IEEE, 2017), pp. 179–184
5. G. Averta, C. Della Santina, E. Battaglia, F. Felici, M. Bianchi, A. Bicchi, Unvealing the principal modes of human upper limb movements through functional analysis. Front. Robot. AI **4**, 37 (2017)
6. G. Averta, F. Angelini, M. Bonilla, M. Bianchi, A. Bicchi, Incrementality and hierarchies in the enrollment of multiple synergies for grasp planning. IEEE Robot. Autom. Lett. **3**, 2686–2693 (2018)
7. G. Averta, G. Valenza, V. Catrambone, F. Barontini, E.P. Scilingo, A. Bicchi, M. Bianchi, On the time-invariance properties of upper limb synergies. IEEE Trans. Neural Syst. Rehabil. Eng. **27**, 1397–1406 (2019)
8. C. Bartneck, D. Kulić, E. Croft, S. Zoghbi, Measurement instruments for the anthropomorphism, animacy, likeability, perceived intelligence, and perceived safety of robots. Int. J. Soc. Robot. **1**(1), 71–81 (2009)
9. E. Battaglia, M. Bianchi, A. Altobelli, G. Grioli, M.G. Catalano, A. Serio, M. Santello, A. Bicchi, ThimbleSense: a fingertip-wearable tactile sensor for grasp analysis. IEEE Trans. Haptics **9**(1), 121–133 (2016)
10. M. Benati, S. Gaglio, P. Morasso, V. Tagliasco, R. Zaccaria, Anthropomorphic robotics. Biol. Cybern. **38**(3), 125–140 (1980)
11. N.A. Bernshteĭn, *The Co-Ordination and Regulation of Movements* (Pergamon Press, Oxford, 1967)
12. M. Bianchi, A. Moscatelli, *Human and Robot Hands: Sensorimotor Synergies to Bridge the Gap Between Neuroscience and Robotics* (Springer, Berlin, 2016)
13. M. Bianchi, G. Averta, E. Battaglia, C. Rosales, M. Bonilla, A. Tondo, M. Poggiani, G. Santaera, S. Ciotti, M.G. Catalano, et al., Touch-based grasp primitives for soft hands: applications to human-to-robot handover tasks and beyond, in *2018 IEEE International Conference on Robotics and Automation (ICRA)* (IEEE, 2018), pp. 7794–7801

14. A. Bicchi, M. Gabiccini, M. Santello, Modelling natural and artificial hands with synergies. Philos. Trans. R. Soc. B **366**(1581), 3153–3161 (2011). https://doi.org/10.1098/rstb.2011.0152
15. M. Bonilla, E. Farnioli, C. Piazza, M. Catalano, G. Grioli, M. Garabini, M. Gabiccini, A. Bicchi, Grasping with soft hands, in *2014 14th IEEE-RAS International Conference on Humanoid Robots (Humanoids)* (IEEE, 2014), pp. 581–587. https://doi.org/10.1109/HUMANOIDS.2014.7041421
16. R.W. Brockett, Robotic manipulators and the product of exponentials formula, in *Mathematical Theory of Networks and Systems* (Springer, Berlin, 1984), pp. 120–129
17. C.Y. Brown, H.H. Asada, Inter-finger coordination and postural synergies in robot hands via mechanical implementation of principal components analysis, in *IEEE/RSJ International Conference on Intelligent Robots and Systems, 2007. IROS 2007* (IEEE, 2007), pp. 2877–2882. https://doi.org/10.1109/iros.2007.4399547
18. S. Casini, V. Tincani, G. Averta, M. Poggiani, C. Della Santina, E. Battaglia, M.G. Catalano, M. Bianchi, G. Grioli, A. Bicchi, Design of an under-actuated wrist based on adaptive synergies, in *2017 IEEE International Conference on Robotics and Automation (ICRA)* (IEEE, 2017), pp. 6679–6686
19. M.G. Catalano, G. Grioli, E. Farnioli, A. Serio, C. Piazza, A. Bicchi, Adaptive synergies for the design and control of the Pisa/IIT softHand. Int. J. Robot. Res. **33**(5), 768–782 (2014). https://doi.org/10.1177/0278364913518998
20. V. Catrambone, A. Greco, G. Averta, M. Bianchi, A. Bicchi, E.P. Scilingo, G. Valenza, EEG complexity maps to characterise brain dynamics during upper limb motor imagery, in *2018 40th Annual International Conference of the IEEE Engineering in Medicine and Biology Society (EMBC)* (IEEE, 2018), pp. 3060–3063
21. V. Catrambone, A. Greco, G. Averta, M. Bianchi, G. Valenza, E.P. Scilingo, Predicting object-mediated gestures from brain activity: an EEG study on gender differences. IEEE Trans. Neural Syst. Rehabil. Eng. **27**(3), 411–418 (2019)
22. W. Chen, C. Xiong, On adaptive grasp with underactuated anthropomorphic hands. J. Bionic Eng. **13**(1), 59–72 (2016). https://doi.org/10.1016/s1672-6529(14)60160-8
23. M. Ciocarlie, C. Goldfeder, P. Allen, Dexterous grasping via eigengrasps: a low-dimensional approach to a high-complexity problem, in *Robotics: Science and Systems Manipulation Workshop-Sensing and Adapting to the Real World* (2007)
24. R. Cubelli, C. Marchetti, G. Boscolo, S. Della Sala, Cognition in action: testing a model of limb apraxia. Brain and Cogn. **44**(2), 144–165 (2000)
25. M.R. Cutkosky, On grasp choice, grasp models, and the design of hands for manufacturing tasks. IEEE Trans. Robot. Autom. **5**(3), 269–279 (1989)
26. A. d'Avella, P. Saltiel, E. Bizzi, Combinations of muscle synergies in the construction of a natural motor behavior. Nat. Neurosci. **6**(3), 300 (2003)
27. C. Della Santina, G. Grioli, M. Catalano, A. Brando, A. Bicchi, Dexterity augmentation on a synergistic hand: the Pisa/IIT softHand+, in *2015 IEEE-RAS 15th International Conference on Humanoid Robots (Humanoids)* (IEEE, 2015), pp. 497–503. https://doi.org/10.1109/humanoids.2015.7363595
28. C. Della Santina, M. Bianchi, G. Averta, S. Ciotti, V. Arapi, S. Fani, E. Battaglia, M.G. Catalano, M. Santello, A. Bicchi, Postural hand synergies during environmental constraint exploitation. Front. Neurorobot. **11**, 41 (2017)
29. C. Della Santina, M. Bianchi, G. Grioli, F. Angelini, M. Catalano, M. Garabini, A. Bicchi, Controlling soft robots: balancing feedback and feedforward elements. IEEE Robot. Autom. Mag. **24**(3), 75–83 (2017)
30. C. Della Santina, C. Piazza, G. Grioli, M. Catalano, A. Bicchi, Towards dexterous manipulation with augmented adaptive synergies: the Pisa/IIT SoftHand 2. IEEE Trans. Robot. **34**, 1141–1156 (2018)
31. C. Della Santina, V. Arapi, G. Averta, F. Damiani, G. Fiore, A. Settimi, M.G. Catalano, D. Bacciu, A. Bicchi, M. Bianchi, Learning from humans how to grasp: a data-driven architecture for autonomous grasping with anthropomorphic soft hands. IEEE Robot. Autom. Lett. **4**(2), 1533–1540 (2019)

32. A. Dragan, S. Srinivasa, Integrating human observer inferences into robot motion planning. Auton. Robots **37**(4), 351–368 (2014)
33. B.R. Duffy, Anthropomorphism and the social robot. Robot. Auton. Syst. **42**(3), 177–190 (2003)
34. C. Eppner, O. Brock, Planning grasp strategies that exploit environmental constraints, in *2015 IEEE International Conference on Robotics and Automation (ICRA)* (IEEE, 2015), pp. 4947–4952. https://doi.org/10.1109/icra.2015.7139886
35. C. Eppner, R. Deimel, J. Álvarez-Ruiz, M. Maertens, O. Brock, Exploitation of environmental constraints in human and robotic grasping. Int. J. Robot. Res. (2015). https://doi.org/10.1177/0278364914559753 https://doi.org/10.1007/978-3-319-28872-7_23
36. T. Feix, J. Romero, H.B. Schmiedmayer, A.M. Dollar, D. Kragic, The grasp taxonomy of human grasp types. IEEE Trans. Hum. Mach. Syst. **46**(1), 66–77 (2016)
37. T. Flash, N. Hogan, The coordination of arm movements: an experimentally confirmed mathematical model. J. Neurosci. **5**(7), 1688–1703 (1985)
38. M. Gabiccini, A. Bicchi, D. Prattichizzo, M. Malvezzi, On the role of hand synergies in the optimal choice of grasping forces. Auton. Robots **31**(2–3), 235–252 (2011)
39. M. Gabiccini, G. Stillfried, H. Marino, M. Bianchi, A data-driven kinematic model of the human hand with soft-tissue artifact compensation mechanism for grasp synergy analysis, in *2013 IEEE/RSJ International Conference on Intelligent Robots and Systems* (IEEE, 2013), pp. 3738–3745
40. S.L. Gorniak, V.M. Zatsiorsky, M.L. Latash, Hierarchies of synergies: an example of two-hand, multi-finger tasks. Exp. Brain Res. **179**(2), 167–180 (2007). https://doi.org/10.1007/s00221-006-0777-z
41. I.V. Grinyagin, E.V. Biryukova, M.A. Maier, Kinematic and dynamic synergies of human precision-grip movements. J. Neurophysiol. **94**(4), 2284–2294 (2005)
42. G. Grioli, M. Catalano, E. Silvestro, S. Tono, A. Bicchi, Adaptive synergies: an approach to the design of under-actuated robotic hands, in *2012 IEEE/RSJ International Conference on Intelligent Robots and Systems (IROS)* (IEEE, 2012), pp. 1251–1256. https://doi.org/10.1109/iros.2012.6385881
43. G. Handjaras, G. Bernardi, F. Benuzzi, P.F. Nichelli, P. Pietrini, E. Ricciardi, A topographical organization for action representation in the human brain. Hum. Brain Mapp. **36**(10), 3832–3844 (2015)
44. K.R. Holzbaur, W.M. Murray, S.L. Delp, A model of the upper extremity for simulating musculoskeletal surgery and analyzing neuromuscular control. Ann. Biomed. Eng. **33**(6), 829–840 (2005)
45. I. Jolliffe, Principal component analysis. Wiley Online Library (2002). https://doi.org/10.1007/978-3-642-04898-2_455
46. M.L. Latash, *Synergy* (Oxford University Press, Oxford, 2008)
47. M.L. Latash, J.P. Scholz, G. Schöner, Toward a new theory of motor synergies. Motor Control **11**(3), 276–308 (2007). https://doi.org/10.1123/mcj.11.3.276
48. J. Lenarcic, A. Umek, Simple model of human arm reachable workspace. IEEE Trans. Syst. Man Cybern. **24**(8), 1239–1246 (1994)
49. M. Malhotra, E. Rombokas, E. Theodorou, E. Todorov, Y. Matsuoka, Reduced dimensionality control for the act hand, in *2012 IEEE International Conference on Robotics and Automation (ICRA)* (IEEE, 2012), pp. 5117–5122. https://doi.org/10.1109/icra.2012.6224651
50. C.R. Mason, J.E. Gomez, T.J. Ebner, Hand synergies during reach-to-grasp. J. Neurophysiol. **86**(6), 2896–2910 (2001)
51. W. Maurel, D. Thalmann, Human shoulder modeling including scapulo-thoracic constraint and joint sinus cones. Comput. Graph. **24**(2), 203–218 (2000)
52. F.A. Mussa-Ivaldi, Modular features of motor control and learning. Curr. Opin. Neurobiol. **9**(6), 713–717 (1999). https://doi.org/10.1016/s0959-4388(99)00029-x
53. A. Naceri, M. Santello, A. Moscatelli, M.O. Ernst, Digit position and force synergies during unconstrained grasping, in *Human and Robot Hands* (Springer, Berlin, 2016), pp. 29–40. https://doi.org/10.1007/978-3-319-26706-7_3

54. J.C. Perry, J. Rosen, S. Burns, Upper-limb powered exoskeleton design. IEEE/ASME Trans. Mechatron. **12**(4), 408 (2007)
55. C. Piazza, C. Della Santina, M. Catalano, G. Grioli, M. Garabini, A. Bicchi, Softhand pro-d: matching dynamic content of natural user commands with hand embodiment for enhanced prosthesis control, in *2016 IEEE International Conference on Robotics and Automation (ICRA)* (IEEE, 2016), pp. 3516–3523
56. C. Piazza, G. Grioli, M. Catalano, A. Bicchi, A century of robotic hands. Annu. Rev. Control Robot. Auton. Syst. **2**, 1–32 (2019)
57. J.O. Ramsay, Functional data analysis. Wiley Online Library (2006)
58. J.O. Ramsay, B.W. Silverman, *Applied Functional Data Analysis: Methods and Case Studies*, vol. 77 (Springer, New York, 2002). Citeseer
59. J.O. Ramsay, G. Hooker, S. Graves, *Functional Data Analysis with R and MATLAB* (Springer Science & Business Media, Berlin, 2009)
60. L.D. Riek, T.C. Rabinowitch, B. Chakrabarti, P. Robinson, How anthropomorphism affects empathy toward robots, in *Proceedings of the 4th ACM/IEEE International Conference on Human Robot Interaction* (ACM, 2009), pp. 245–246
61. P. Saltiel, K. Wyler-Duda, A. D'Avella, M.C. Tresch, E. Bizzi, Muscle synergies encoded within the spinal cord: evidence from focal intraspinal NMDA iontophoresis in the frog. J. Neurophysiol. **85**(2), 605–619 (2001)
62. M. Santello, M. Flanders, J.F. Soechting, Postural hand synergies for tool use. J. Neurosci. **18**(23), 10105–10115 (1998)
63. M. Santello, M. Flanders, J.F. Soechting, Patterns of hand motion during grasping and the influence of sensory guidance. J. Neurosci. **22**(4), 1426–1435 (2002)
64. M. Santello, G. Baud-Bovy, H. Jörntell, Neural bases of hand synergies. Front. Comput. Neurosci. **7**, 23 (2013)
65. M. Santello, M. Bianchi, M. Gabiccini, E. Ricciardi, G. Salvietti, D. Prattichizzo, M. Ernst, A. Moscatelli, H. Jörntell, A.M. Kappers, et al., Hand synergies: integration of robotics and neuroscience for understanding the control of biological and artificial hands. Phys. Life Rev. **17**, 1–23 (2016). https://doi.org/10.1016/j.plrev.2016.02.001
66. R.W. Schafer, What is a Savitzky-Golay filter? [lecture notes]. IEEE Signal Process. Mag. **28**(4), 111–117 (2011). https://doi.org/10.1109/msp.2011.941097
67. J.P. Scholz, G. Schöner, The uncontrolled manifold concept: identifying control variables for a functional task. Exp. Brain Res. **126**(3), 289–306 (1999). https://doi.org/10.1007/s002210050738
68. A. Schwarz, G. Averta, J.M. Veerbeek, A.R. Luft, J.P. Held, G. Valenza, A. Biechi, M. Bianchi, A functional analysis-based approach to quantify upper limb impairment level in chronic stroke patients: a pilot study, in *2019 41st Annual International Conference of the IEEE Engineering in Medicine and Biology Society (EMBC)* (IEEE, 2019), pp. 4198–4204
69. P. Stratmann, D. Lakatos, A. Albu-Schäffer, Neuromodulation and synaptic plasticity for the control of fast periodic movement: energy efficiency in coupled compliant joints via PCA. Front. Neurorobot. **10** (2016). https://doi.org/10.3389/fnbot.2016.00002
70. P.H. Thakur, A.J. Bastian, S.S. Hsiao, Multidigit movement synergies of the human hand in an unconstrained haptic exploration task. J. Neurosci. **28**(6), 1271–1281 (2008). https://doi.org/10.1523/jneurosci.4512-07.2008
71. E. Todorov, Z. Ghahramani, Analysis of the synergies underlying complex hand manipulation, in *Engineering in Medicine and Biology Society, 2004. IEMBS'04. 26th Annual International Conference of the IEEE*, vol. 2 (IEEE, 2004), pp. 4637–4640
72. M.C. Tresch, A. Jarc, The case for and against muscle synergies. Curr. Opin. Neurobiol. **19**(6), 601–607 (2009)
73. M.C. Tresch, P. Saltiel, E. Bizzi, The construction of movement by the spinal cord. Nat. Neurosci. **2**(2), 162–167 (1999)
74. M.T. Turvey, Action and perception at the level of synergies. Hum. Mov. Sci. **26**(4), 657–697 (2007)

75. F.J. Valero-Cuevas, Predictive modulation of muscle coordination pattern magnitude scales fingertip force magnitude over the voluntary range. J. Neurophysiol. **83**(3), 1469–1479 (2000)
76. L. Villani, F. Ficuciello, V. Lippiello, G. Palli, F. Ruggiero, B. Siciliano, Grasping and control of multi-fingered hands, in *Advanced Bimanual Manipulation* (Springer, 2012), pp. 219–266. https://doi.org/10.1007/978-3-642-29041-1_5
77. R. Vinjamuri, M. Sun, C.C. Chang, H.N. Lee, R.J. Sclabassi, Z.H. Mao, Temporal postural synergies of the hand in rapid grasping tasks. IEEE Trans. Inf. Technol. Biomed. **14**(4), 986–994 (2010)
78. K. Xu, H. Liu, Y. Du, X. Zhu, Design of an underactuated anthropomorphic hand with mechanically implemented postural synergies. Adv. Robot. **28**(21), 1459–1474 (2014)

Learning from the Human Hand: Force Control and Perception Using a Soft-Synergy Prosthetic Hand and Noninvasive Haptic Feedback

Qiushi Fu and Marco Santello

Abstract Force control and perception plays an important role in activities of daily living when handling objects with different physical properties. These abilities are results of complex sensorimotor pathways that coordinate movements, predict consequences, and process feedback. For prosthetic systems, the ability to exhibit human-like action and perception behavior is critical for the acceptance of the terminal device. In this chapter, we review recent findings obtained from a bioinspired soft-synergy prosthetic hand and a noninvasive mechanotactile feedback device. A series of experiments demonstrated the improvement in force control and perception in closed-loop prosthesis through context-aware myoelectric controllers and contralateral haptic training protocols. By comparing performances between human native hands and prosthetic hands, we provide novel insights on the importance of learning from human sensorimotor mechanisms in the design of upper-limb neuroprosthesis.

Keywords Neuroprosthetics · Haptic feedback · Force control · Perception

1 Introduction

Restoring the ability to interact with the environment through prosthetic hands is critically important for individuals with upper-limb loss to regain independence in activities of daily living. However, despite decades of advances in prosthetic research, significant limitations in the reliability, function, and robustness in prosthetic hands still exist to prevent human-level dexterity. Recently, human-

Q. Fu (✉)
Department of Mechanical and Aerospace Engineering, University of Central Florida, Orlando, FL, USA
e-mail: Qiushi.Fu@ucf.edu

M. Santello
School of Biological and Health Systems Engineering, Arizona State University, Tempe, AZ, USA

R. Vinjamuri (ed.), *Advances in Motor Neuroprostheses*,
https://doi.org/10.1007/978-3-030-38740-2_4

inspired approaches have been proposed to address these challenges through novel mechanical design [1], intuitive control [2, 3], and artificial sensory feedback [4]. Specifically, it was observed that the central nervous system (CNS) solves the sensorimotor complexity of grasp control through neuromuscular synergies (i.e., consistent joint/muscle coordination patterns [5–7]), which enables the control of a large number of muscles and joints through a smaller set of neural signals [for review see 8]. The combination of synergies with soft robotic technologies has inspired the creation of compliant and underactuated prosthetic hands that simultaneously maximize simplicity and functionality [9]. Furthermore, it is well-known that somatosensory information plays key role during the generation of interaction forces in skilled manual functions [10]. As the central nerves system faces the challenge of noise and uncertainty, somatosensory feedback through tactile and proprioceptive sensory channels can be used not only to generate reactive motor commands when execution errors occur but also to facilitate future predictive actions by updating the internal representation of the hand-object interactions [11]. Therefore, providing somatosensory feedback (i.e., close the loop) should benefit users of prosthetic hands and improve the usability of the prosthetic systems, especially in situations where visual feedback is missing (e.g., the hand prosthesis is out of view) or ineffective (e.g., force control). Many different forms of sensory feedback have been implemented and tested in hand prostheses [12]. One of the noninvasive approaches is modality-matched feedback, which delivers stimuli using sensory channels of same modality as the missing ones, although the location of the missing physiological feedback may not be matched, e.g., pressure sensed via a prosthetic hand is delivered through pressure on forearm. In this chapter, we review important new findings obtained by research on biologically inspired, closed-loop prosthetic hands that demonstrated human-like behavior in the control and perception of grasp forces.

2 Materials and Method

2.1 Subjects

Sixteen (9 females and 7 males, ages 19–34) and eighteen volunteers (9 males and 9 females, ages 18–37) participated in the first and second study, respectively. All subjects were right-handed according to the Edinburgh Handedness Questionnaire. They had normal or corrected-to-normal vision, no previous history of orthopedic, neurological trauma, or pathology of the upper limbs and were naive to the purpose of the study. Subjects gave their informed consent, and the protocols were approved by the Institutional Review Board at Arizona State University in accordance with the Declaration of Helsinki.

2.2 Experiment Apparatus

For the present investigations of force control and perception, we used a closed-loop prosthetic system that consists of a myoelectric prosthetic hand (right hand) with customized socket interface and a mechanotactile feedback device (Fig. 1). Additionally, a gravity compensation system was used to off-load the added weight which was caused by able-bodied subjects wearing the prosthetic hands. Lastly, we used customized objects with embedded force/torque sensors to measure subjects' grasp forces during different grasping tasks. We will briefly describe each component of this experiment setup below (for more details, the reader is referred to [13]).

2.2.1 SoftHand-Pro (SHP)

The SHP is the prosthetic version of the Pisa/IIT SoftHand [14]. The size and weight of the SHP approximate a large male hand, and a glove was used to protect the cables and increase contact area and friction during hand-object interactions. It has been shown that transradial amputees are able to use the SHP effectively in activities of daily living [15]. The core feature of the SHP is that it only uses one DC motor to drive all finger joints synergistically based on the first hand postural synergy (first principal component described by [5]). The grasping behavior of the SHP is facilitated by embedded joint compliance to mechanically adapt to different object shapes. The single motor is controlled by a PID position/current controller. Two surface EMG electrodes (13E200 Myobock electrodes, Otto Bock, Germany) located over wrist flexor and extensor muscles were used to control the closing and opening of fingers, respectively. The SHP does not directly sense contact forces from the fingers but rather estimates the overall grasp force from the motor current [2]. The grasp force estimation is used for driving the haptic feedback device, as well as for determining the operating state of the prosthesis.

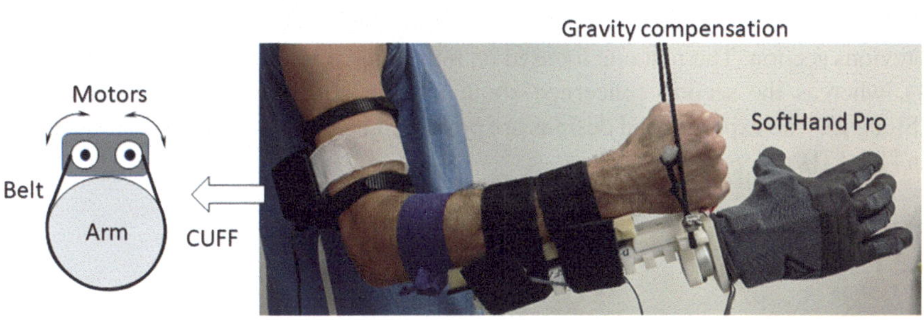

Fig. 1 Components of the implemented closed-loop prosthetic system. (Adapted from [13])

2.2.2 Clenching Upper-Limb Force Feedback Device (CUFF)

The CUFF consists of two DC motors that tighten or loosen a fabric belt worn around the upper arm. Therefore, pressure can be generated around the arm proportional to the estimated grasp force from the SHP [16]. This type of mechanotactile stimuli matches the modality of somatosensation involved in natural grasping and manipulation, although it is delivered to a different location. The CUFF was calibrated for each subject to accommodate the individual differences in arm size and softness.

2.2.3 Gravity Compensation

A light cable, a series of pulleys, and a counterweight were used to support the weight of the prosthetic hand and the socket interface. Our experiments required intensive repetition of grasp force production with the entire arm held in air. Therefore, gravity compensation system was required to prevent muscle fatigue that may arise from sustaining the weight of the prosthetic system during long-time use.

2.2.4 Data Recording

The kinematics of the object and hands was tracked using a motion capture system (Impulse, PhaseSpace Inc.) with markers placed on the objects and the wrist. We instrumented sensorized objects with different size and weight. Each object is equipped with a force/torque sensor (Nano 25, ATI Industrial Automation, NC, USA) at the center to record the resultant grasp force.

2.3 *Experimental Designs*

We conducted two studies using the closed-loop prosthetic system described in the previous section. The first one focused on improving fine grasp force control (Study 1), whereas the second concerned about inter-limb transfer of force perception (Study 2). The experimental designs are presented below (more details can be found in [13, 17]).

2.3.1 Study 1

In this study, we compared two myoelectric controllers (single-gain and hybrid-gain) using tasks that required subjects to grasp and transport objects with different physical properties (i.e., size, weight, and fragility). Both controllers have the same EMG processing component but differ in the EMG-to-motion mapping algorithms

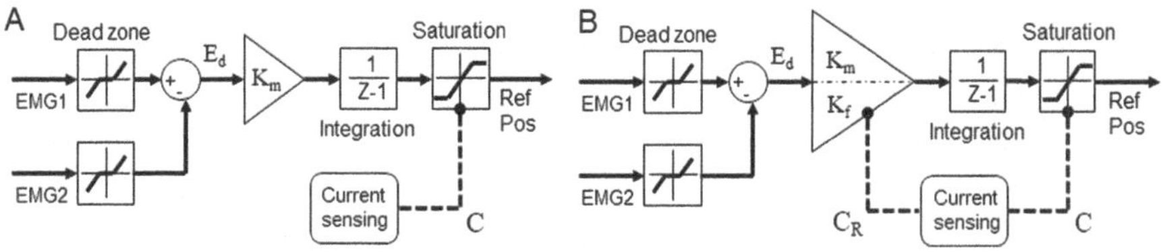

Fig. 2 Block diagrams of the single-gain (Panel **a**) and hybrid-gain (Panel **b**) myoelectric control. (Adapted from [13])

(Fig. 2). Specifically, a signal dead zone of 2% MVC sEMG signal was first applied to each channel, then the channel difference E_d between the sEMG signals from flexor and extensor muscles was used to drive the change of SHP motor reference position with predetermined gain(s). Note that the grasp force is generated due to the PID position controller trying to follow the reference position while being stopped by the hand-object contact. An adaptive motor position limit was implemented to prevent the increase of reference position if the motor total current C is close to the max capacity, thus allowing consistent opening motion from objects with any size.

The single-gain (SG) controller is mostly identical to the best performing SHP motion controller demonstrated by Fani and colleagues [18], in which a constant gain K_m was used (Fig. 2a). The main drawback of the SG controller is that it cannot adapt to both free motion control and grasp force control equally well. A gain optimized for motion control is usually much larger for the optimal gain for fine force control. To overcome this problem, we created a hybrid-gain (HG) controller in which the EMG-to-motion gain changes adaptively depending on the state of the SHP (Fig. 2b). Three sensorimotor states of the SHP were defined using the residual current C_R as well as the EMG differential E_d. Specifically, free motion state is when the grasp force is zero or very low, i.e., $C_R = 0$. Fine force state is when the grasp force is above minimum and the user is trying to control grasp force, i.e., $C_R > 0$ and $E_d > 0$. The last state, quick release, is when the grasp force is above minimum and the subject is trying to quickly release the grasped object, i.e., $C_R > 0$ and $E_d < 0$. We used a large gain K_m for both free motion and quick release states and a small gain K_f for fine force state. Unlike previous work [2], the control gain for the internal motor control loop remain unchanged, therefore preserving the stability during the passage from one state to another.

Sixteen subjects were equally assigned to two groups that used either SG or HG controllers. Both groups had the same experimental protocol that consisted of three sessions: (1) training trials, training tasks with SHP and CUFF; (2) baseline trials, experimental tasks with native right hand; and (3) SHP trials, experimental tasks with SHP and CUFF. Baseline trials were performed first to acquire data from native hand as a benchmark to evaluate the performance of the prosthetic system.

The training sessions implemented two tasks that help subjects to familiarize with myoelectric control of the SHP and haptic feedback from the CUFF. First, the motion control training task required subjects to control the open and close of the

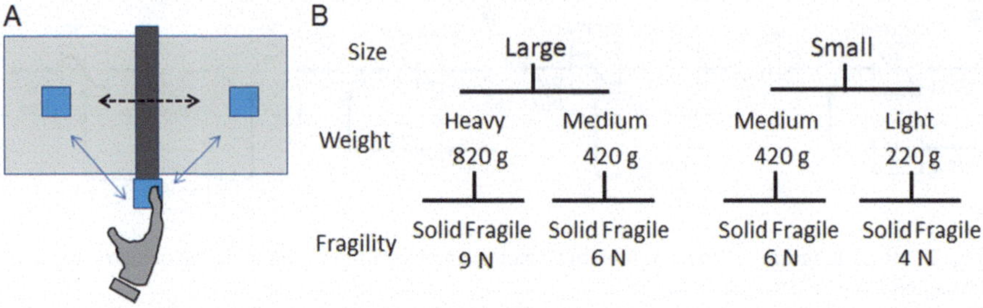

Fig. 3 Experimental tasks used in Study 1. Panel **a** shows the top view of the area of movement, and Panel **b** shows the object properties used in the study. (Adapted from [13])

SHP to match the target motor positions as quickly as possible, following visual feedback on a monitor. Six levels of target motor position were used: 0°, 30°, 60°, 90°, 120°, and 150°. Second, the force control training task required subjects to close the SHP on a cylindrical object to match the different levels of target grasp forces. Three levels of target force were defined: 0 N, 6 N, and 12 N. In addition to visual feedback, subjects also received haptic feedback from the CUFF.

Both the baseline and SHP trials used the same experimental tasks. These tasks were inspired by commonly used clinical hand function assessment tools (e.g., Southampton Hand Assessment Procedure, Block and Box Test, etc.), with the focus on the ability of fine control of grasp forces during functional use of the prosthetic hand. Specifically, subjects were instructed to pick and place sensorized objects repetitively between two target regions separated by 30 cm (Fig. 3a). A 5-cm high metal bar was placed on the midline between the two target regions as an obstacle, and subjects were required to move back to the proximal end of the bar after each successful pick and place action. Success was defined as not dropping or "crushing" the object during transport. The crushing of the object was rendered by giving "glass breaking" sound when the grasp force exceeded a pre-defined crushing threshold based on its fragility. The target objects varied in size, weight, and fragility (Fig. 3b), and for each object, subjects were told to complete pick and place actions successfully as many time as possible within 45 s. Note that large and small objects require power and precision grasps, respectively. Despite of the grasp type, we aimed to examine the performance of SG and HG controllers for the modulation of grasp forces in response to object weight and fragility in this study.

To quantify subjects' performance in the experimental tasks, we computed the following variables: (a) number of successful transport completed within 45 s and (b) grasp force during successful transport.

2.3.2 Study 2

In this study, we compared two force training protocols (ipsilateral and contralateral training) using tasks that required subjects to reproduce a grasp force with one

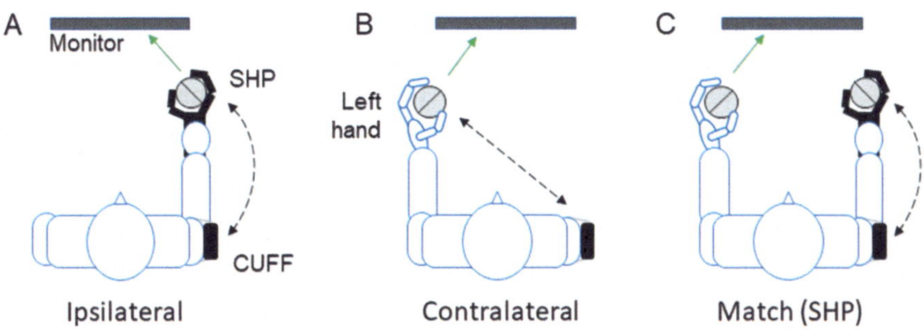

Fig. 4 Panels **a** and **b** show the design of ipsilateral and contralateral training tasks, respectively. Panel **c** shows the design of the force matching task. Solid arrows represent visual feedback, and the black dashed arrows represent haptic feedback. (Adapted from [17])

hand that matches the preceding force generated by the contralateral hand. Note that the myoelectric interface used here is the hybrid-gain controller developed in Study 1. The ipsilateral training was similar to the force training used in Study 1, allowing subjects to establish the association between the SHP grasp force and CUFF pressure feedback (Fig. 4a). In contrast, the contralateral training was designed to help subjects learn the mapping between the CUFF pressure and the natural force perception from their native left hand (Fig. 4b). Specifically, subjects grasp the sensorized cylinder with their left hand to produce difference force levels following visual feedback. The grasp force measured from the cylinder was mapped linearly to the CUFF motor, such that the same CUFF pressure represented the same measured grasp force in both contralateral and ipsilateral training tasks, i.e., net force from fingers. For both training protocol, four force targets were used: 0 N, 4 N, 6 N, 8 N, and 10 N.

The main experimental task for this study is a contralateral force matching task. Subjects were instructed to first maintain a target grasp force within $\pm 5\%$ of the target with their left hand (i.e., reference hand) for a cumulative duration of 6 s. Immediately after releasing the left hand, subjects generated grasp force with their right hand (native or SHP) to match the memorized force level (Fig. 4c). The CUFF was always paired with the SHP delivering feedback about SHP grasp force. Note that subjects only had partial visual feedback when using their left hand, which displayed the relative strength of the grasp force (i.e., percentage of target) as a moving bar, without giving information about the actual force magnitude. There were three force targets: 4 N, 7 N, and 10 N. During force matching, subjects were instructed to give a verbal confirmation when they felt the force they were generating with the right hand (native or SHP) matched the target, and the experimenter recorded the confirmation time. Note that the force matching task with the SHP-CUFF system requires subjects to have learned the mapping between myoelectric control, grasp force, and CUFF pressure acquired through ipsilateral and contralateral training. Furthermore, the 4 N and 10 N force levels were included in the training session, but the 7 N force level was not. This was designed to examine the extent of training generalization.

Eighteen subjects were equally assigned to two groups that received different force training combinations that consisted of mixtures of ipsilateral and contralateral tasks with different proportions. There were four blocks of training trials for each group. Within each block, the EMG-haptic group started with one contralateral training trial, followed by three ipsilateral training trials. In contrast, the native-haptic group started with three contralateral training trials followed by one ipsilateral training trial per block. Before and after the training session, subjects performed native hand and SHP matching tasks, respectively. For either one of these matching tasks, there were a total of 24 trials with each force matching target appearing 8 times in random order.

We computed the following two variables to quantify the performance of force matching: (a) *Relative matching error*, defined as the difference between the force produced by the reference hand (averaged across a 6-s window before release) and matching hand (averaged across a 0.6-s window before subject's confirmation). A positive or a negative relative error means subjects generated more or less force with the right hand (native or SHP) than the left hand during matching, respectively. (b) *Absolute matching error* was defined as the absolute value of the relative errors, which can be considered as a measurement of variability if no bias is found in the relative error.

3 Study 1 Results: Fine Control of Grasping Force During Hand-Object Interactions

The complete results of this study can be found in [13]. Here we only report two main findings regarding the performance of the pick-and-place task. As a benchmark, the three-way mixed ANOVA (group × weight × fragility) was used for each object size to compare the number of successful transport within 45 s for trials performed by the native hands. It was found that subjects from SG and HG groups performed equally well. Furthermore, only fragility but not weight of the objects had influenced the net performance (Fig. 5a, b). The number of successful transport for fragile objects was significantly less than the transport of the solid ones (only main effect of fragility with both large and small object $p < 0.001$). This indicated that subjects handled the fragile objects using slower speed with more caution. While using SHP, both groups performed the tasks much slower than their native hands, and we found that the HG controller outperformed the SG controller when transporting fragile objects (Fig. 5c, d). The three-way mixed ANOVA (group, weight, and fragility) for large objects revealed significant fragility × group ($p = 0.003$) and fragility × weight interactions ($p = 0.023$). Post hoc comparisons suggested that the HG group performed significantly better than the SG group in heavy-fragile, medium-solid, and medium-fragile conditions ($p < 0.05$; Fig. 5c). No difference was found between the two groups in the heavy-solid condition. Similarly with the small object, we found a significant fragility × group interaction

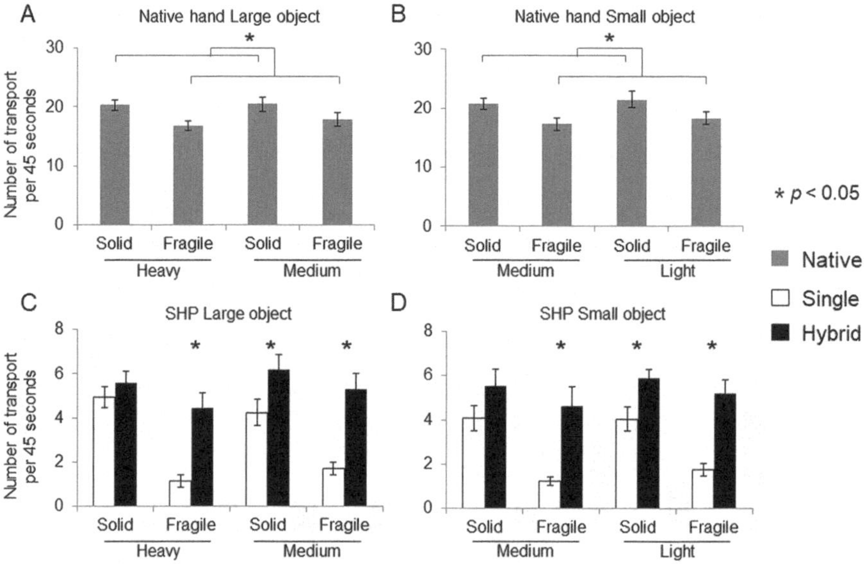

Fig. 5 Number of successful pick-and-place within 45 s (Mean ± SE). (Adapted from [13])

($p = 0.035$). Further, t-test suggested that the hybrid group performed significantly better than the differential group in heavy-fragile, medium-solid, and medium-fragile conditions ($p < 0.05$; Fig. 5d), but not in heavy-solid condition. Importantly, we demonstrated a qualitatively similar pattern of fragility effect between the SHP and native hand in the HG group but not the SG group, despite the significantly lower number of successful trials overall.

As for the grasp forces, we found that subjects scaled grasp force to object weight and fragility in both object size conditions when using their native hands (Fig. 6a, b). Specifically, subjects used larger grasp force for heavier objects, and smaller grasp force when the object was fragile. These observations were confirmed by the three-way mixed ANOVA (group, weight, and fragility). With the large object, there was a significant main effect of both weight ($p = 0.003$) and fragility ($p < 0.001$), but not group. Similarly with the small object, we found a significant main effect of both weight ($p < 0.001$) and fragility ($p < 0.001$), but not group. With the SHP, subjects were able to modulate the grasp force in a successful transport (Fig. 6c, d). With the large object, we found a main effect of weight ($p < 0.001$) and a significant fragility × group interaction ($p = 0.024$). Post hoc comparisons showed that the HG group used a significantly smaller grasp force than the SG group in medium-solid condition ($p < 0.05$; Fig. 6c). Similarly with the small object, we also found a main effect of weight ($p = 0.003$) and a significant fragility × group interaction ($p < 0.001$). T-test showed that the HG group used a significantly smaller grasp force than the SG group in both medium-solid and light-solid conditions ($p < 0.05$; Fig. 6d). When compared with native hand, we found that the HG group showed a qualitative similar pattern of grasp force modulation, but the SG group did not.

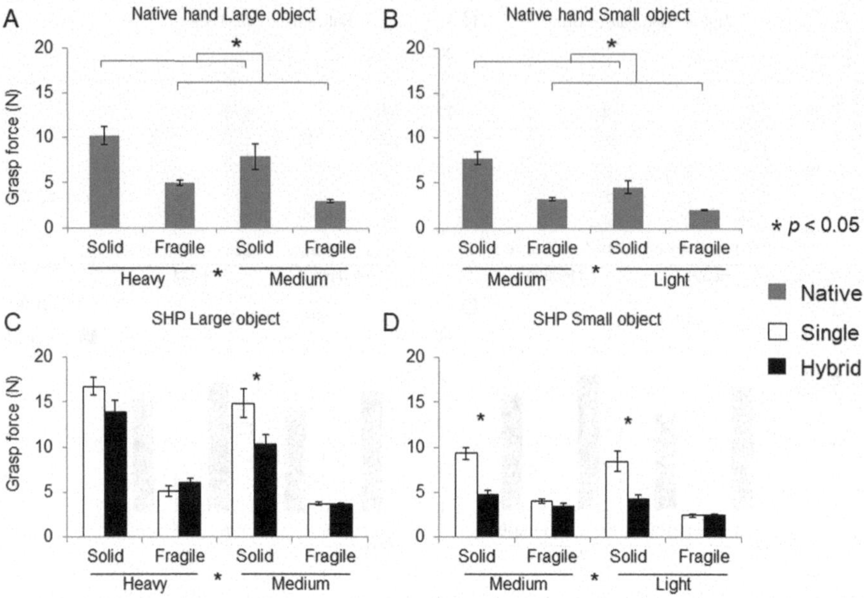

Fig. 6 Averaged grasp force during a successful transport (Mean ± SE). (Adapted from [13])

4 Study 2 Results: Inter-Limb Transfer of Perceptual Information About Grasping Force

The complete results of this study can be found in [17]. Here we only report the main findings regarding the performance contralateral force matching task. To determine what trials should be used to evaluate matching accuracy, we first assessed whether there was a general learning effect through multiple trials. This was accomplished by averaging the absolute error across three force levels for each of the eight trials. We performed the two-way mixed ANOVA (trial × group) for both the native hand and SHP matching sessions, and both sessions exhibited a significant main effect of trial (native hand: $p < 0.001$; SHP: $p < 0.001$). After visual inspection of the trends (Fig. 7a), we decided to assess the interaction between training protocols and force levels (group × force) by averaging the last five trials which exhibited performance plateau. For the native hand, we found a significant effect of force ($p = 0.003$). Post hoc comparisons revealed that subjects were less accurate in the 10 N condition than 4 N and 7 N conditions (Fig. 7b). For the SHP, neither group nor force effect was significant. Overall, these results suggest that subjects tended to perform the best in the 7 N condition with their native right hand, but worst with the SHP/CUFF system. Furthermore, the two training protocols did not lead to the differences in the magnitude of matching errors.

We also averaged the relative errors across the last five trials per force level based on the above results. We performed the two-way mixed ANOVA (group × force)

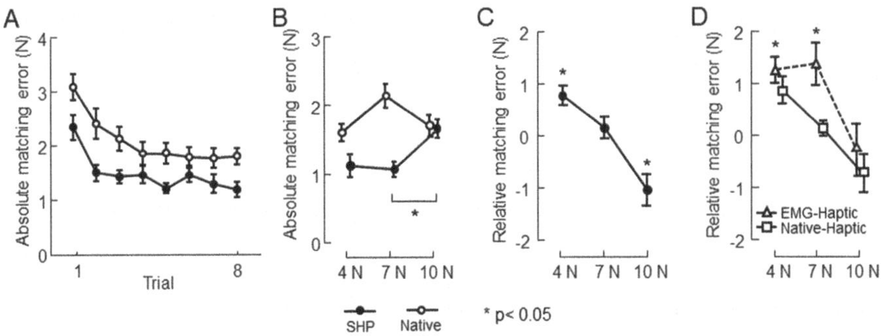

Fig. 7 Experimental result from force matching tasks (mean ± SE). (Adapted from [17])

for the native hand matching and SHP matching sessions separately. For the native hand, we found only a significant effect of force ($p < 0.001$). One sample t-test revealed that subjects made positive errors in the 4 N conditions (0.77 ± 0.19 N), neutral errors in the 7 N conditions (0.14 ± 0.21 N), and negative errors in the 10 N conditions (−1.06 ± 0.30 N; Fig. 7c). For the SHP, we found significant effects of both force ($p = 0.002$) and group ($p = 0.043$). One sample t-test revealed that subjects in the EMG-haptic group made positive errors in both 4 N and 7 N conditions (1.25 ± 0.25 N and 1.36 ± 0.41 N) and neutral errors in 10 N conditions (−0.29 ± 0.50 N). In contrast, the native-haptic group made neutral errors in all three force conditions (0.70 ± 0.34 N, 0.07 ± 0.36 N, and −0.48 ± 0.30 N for 4 N, 7 N, and 10 N, respectively; Fig. 7d). In summary, we found that the native-haptic training helped subjects to generate matching performance in SHP matching qualitatively similar to native hand matching, especially in 7 N conditions. In contrast, the EMG-haptic training led to a significant positive bias for the SHP matching in 7 N conditions.

5 Discussion

This chapter reviewed two recent findings regarding force control and perception in a novel closed-loop hand prosthesis. Specifically, it was found that (1) a context-dependent myoelectric controller could significantly improve the subjects' ability to perform tasks that require fine force control (i.e., transporting objects with different physical properties) and (2) training on the association between the force perception of the intact limb and the artificial haptic feedback of the prosthesis significantly improved inter-limb transfer of force perception at untrained force levels. Importantly, for both studies we compared the performance of the prosthetic system with benchmark data obtained with subjects' intact normal hands. We propose that it is critical to equip prosthetic systems with human-like kinematic and kinetic behavior for increasing the acceptance of the terminal devices. In the following sections, we discuss these results in relation to the putative neural principles underlying human control and perception of forces, as well as the implications to future development of next-generation motor neuroprosthesis.

5.1 Context-Dependent Hybrid Gain Myoelectric Controller

The ability to control grasp force precisely is an important feature in human manual dexterity. For example, it has been shown that the grasp force is often regulated with respect to the object's weight and friction to maintain a "safety margin," i.e., the applied grasp force is slightly higher than the minimally required force such that energy efficiency and slip prevention are balanced [19, 20]. As demonstrated in our study, when friction is constant, such neural control led to the natural scaling of grasp force to match the object weight (e.g., heavier objects were grasped with a larger force normal to the object surface). An extensive investigation has revealed that weight-specific grasp force scaling can be explained by a combination of memory-based predictive control and corrections driven by sensory feedback. Specifically, a feedforward motor command can be generated to interact with a novel object according to the object's physical properties indicated by visual cues during the initial encounter [21]. If there is mismatches between the predicted and actual outcome of the hand-object interactions (indicated by somatosensory feedback), corrective responses are generated [22, 23]. Furthermore, internal representation of the object properties can be updated through a repetitive interaction with the same object, such that the future feedforward motor command can be generated more precisely [11]. Importantly, the feedforward adjustment of "safety margin" for grasp force can also be made according to contextual cues to account for uncertainty in the dynamic environment [24] or the fragility of objects [25]. Our results were consistent with these previous studies. Subjects generated less grasp force when grasping the fragile object than the solid ones of the same weight with their native hands. Such drop of "safety margin" was also accompanied by a decrease in arm movement speed to reduce the force variability induced by dynamic motion [13].

Grasp force control in prosthetic hands can be enabled in several ways. One approach is to implement a feedback control loop within the terminal device to allow automated adjustment using force and position sensors [26–28]. This method could enable good precision and reliability in force generation for a single degree-of-freedom rigid prosthetic hand (e.g., motion control hand). However, replication of this approach in multi-finger hands and/or hands with embedded compliance (e.g., SHP) can be very difficult, because of the need to measure joint kinematics and contact forces accurately. Alternatively, users can have a full control over the grasp force. In these scenarios, haptic feedback is often provided, which allows the subjects to adjust the grasp force accordingly [2, 4, 29, 30]. However, this approach can potentially increase the complexity of the myoelectric interface, leading to a higher cognitive load and attentional demands required to simultaneously control multiple variables. A middle ground can be found using a shared control scheme in which both the user and the prosthetic system contribute to force control [31]. Following this idea, there have been several recent studies that explored the myoelectric controller designs that switch control schemes based on sensory or motor information [32–36]. It has been argued that sharing control between the user and the terminal device can shield low-level execution details, thus decreasing cognitive burden [37]. This assessment was extended in our study by merging both the sensory and motor streams to determine the context in which the prosthesis operate. Specifically, our novel controller changes behavior based on both the state of the hand (i.e., free motion or object in hand) and the user intent (i.e., grasp or release). Furthermore, we propose that, in a proportional EMG control scheme with a soft robotic hand such as the SHP, the EMG-to-motor control mapping needs to be separately optimized for motion and force. Such implementation is important to allow users to accurately generate both feedforward motor commands and fine corrective motor responses, both of which are crucial to human's manual dexterity. Taken together, our bioinspired context-aware myoelectric controller successfully demonstrated human-like behavior in prosthetic hands.

We should note that the superior performance of the hybrid gain controller was accompanied by increased energy cost as denoted by greater sEMG amplitude and longer contractions [13]. However, this is not necessarily a drawback of the proposed approach. The fact that subjects scaled grasp force to object weight, using the prosthesis in a way similar to using their own hands, indicates that the increased energy cost could effectively evoke the CNS's tendency of optimizing motor commands for energy efficiency. This could lead to a lower energy consumption in the prosthetic hand as the grasp force is optimized, which can subsequently extend the usage time because of less battery usage and reduced tension in the driving tendon.

5.2 Inter-Limb Transfer of Perceptual Information in Closed-Loop Prosthetic Systems

Mechanisms of force perception have often been investigated using asynchronized force matching paradigms (i.e., the force of the reference hand is first memorized). The errors in this task with the native hands can be mostly attributed to errors in perception, assuming the force control in the matching hand is accurate and the memory of the reference force is undergoing little decay in time. The leading theory of force perception is that both peripheral (afferent) and central (efferent) information are integrated [38]. The peripheral afferent signals include sensory receptor discharges arising from Golgi tendon organs during active muscle contractions, as well as discharges of mechanoreceptors in the skin that respond to the different types of the haptic stimuli (e.g., pressure and stretch). As for the central signals, copies of outgoing motor commands (i.e., efferent copy or corollary discharge [39]) are used to predict the incoming sensory information through the feedforward internal models [40]. It was shown that contralateral force matching can still be performed with reasonable accuracy in a deafferent patient [41]. Furthermore, if nonhomologous muscle groups are used in producing perceptually equivalent forces, muscles with larger force capabilities tend to exhibit a larger absolute sEMG amplitude [42]. Therefore, it has been argued that the efferent copies play a major role in force perception during active force generation (e.g., left hand in Study 2). However, peripheral information also contribute in active force perception, since it has been shown that constraining tactile feedback could increase the force matching error [43]. Furthermore, peripheral signals such as tactile feedback play a major role in passive force perception when efferent copy is not available, as indicated by high accuracy in ipsilateral passive finger force matching [44].

Force matching using the prosthetic hand as the matching hand in this study presents new challenges to the CNS, since the control and sensing of the prosthetic system significantly differ from the native hand. Specifically, the native hands generate and maintain the grasp forces mainly through contraction of the digit flexors, and the magnitude of the grasp force is approximately proportional to the magnitude of the EMG activity. In contrast, for prosthetic hand control, contraction of a different muscle group (i.e., wrist flexors) was mainly used. More importantly, the sEMG was mapped to the change of the grasp force instead of the force magnitude. Therefore maintaining grasp force requires the muscles to be at rest instead of contraction. This type of velocity-based myoelectric control interface is commonly used in prosthetic hands, because the control signal can be more stable than those generated from direct mapping of the magnitude (i.e., position-based). However, the sharp differences between velocity-based myoelectric control and the natural neuromuscular control in force generation suggest that the CNS likely does not possess the ability to make precise sensory predictions using ongoing motor commands. From the perspective of sensing using the prosthetic system, subjects can only receive tactile feedback at a location incongruent with the reference hand (i.e., upper arm). Furthermore, the skin of the arm has a much smaller mechanoreceptor density than the fingertips of the native hand [45]. In addition to this spatial mismatch, subjects do not receive grasp force feedback in-sync with muscle contraction due to the use of a velocity-based myoelectric interface in the prosthetic system. In contrast, the native hand can integrate sensory information about the grasp force temporally aligned with the motor command.

Based on the above considerations, the inferior matching performance by the prosthetic hand can be attributed to its non-physiological design features. Most importantly, the force perception using the prosthetic system cannot rely on the "efferent copy" information due the velocity-based control interface. Therefore, the perceptual information from the matching prosthetic hand had to mostly rely on the noisy pressure signal at the CUFF-skin interface, which results in a larger trial-to-trial variability, i.e., absolute error. Moreover, the low proficiency of the force control and the inherent noise and delay in estimating grasp force using motor current of the SHP could make the matching judgment more challenging, leading to longer matching times than those from the native hands [17].

Inter-limb transfer of perceptual information between the intact and prosthetic limb in the force matching tasks can be improved through appropriate training. Specifically, the contralateral training was designed to help subjects experience concurrent perceptual information from both limbs, which could help subjects generate more accurate and reliable perceptual mapping between the two sides. This benefit was demonstrated in the native-haptic group who had no bias in producing matching force at the untrained force level. The observed generalization strongly suggests that inter-limb association of perceptual information can be established between an intact limb and a closed-loop prosthesis through appropriate training paradigm, despite differences in the underlying neural mechanisms.

5.3 Open Questions and Future Research

Our understanding of the factors that limits the performance and acceptance of motor neuroprosthesis has improved significantly over the past two decades. Many researchers have recognized that the integration of sensory feedback in closed-loop hand prosthesis could be beneficial. However, the optimal design of these systems remains to be investigated. One of the important insights we obtained through the studies presented in this chapter is the complementary role of the myoelectric interface in force control and perception. When designing myoelectric interfaces for a closed-loop prosthesis, we propose that more physiologically plausible algorithms should be considered and in particular algorithms that enable the CNS to utilize existing sensorimotor pathways. Furthermore, a context-aware controller should be preferred to assist the user in a wide range of different hand-object interaction scenarios, with their specific optimal gains and state transitions. Lastly, future research should aim at revealing the neural mechanisms underlying the learning processes underlying sensorimotor integration during use of closed-loop neuroprosthesis, which would greatly improve the design of training protocols of prosthetic users.

Acknowledgments This work was made possible by grant W911NF-17-1-0049 from the Defense Advanced Research Projects Agency and the US Army Research Office.

References

1. S.B. Godfrey, A. Ajoudani, M.G. Catalano, G. Grioli, A. Bicchi, A synergy-driven approach to a myoelectric hand. IEEE Int. Conf. Rehabil. Robot. **2013**, 1–6 (2013)
2. A. Ajoudani et al., Exploring teleimpedance and tactile feedback for intuitive control of the Pisa/IIT SoftHand. IEEE Trans. Haptics **7**(2), 203–215 (2014)
3. N. Jiang, H. Rehbaum, I. Vujaklija, B. Graimann, D. Farina, Intuitive, online, simultaneous, and proportional myoelectric control over two degrees-of-freedom in upper limb amputees. IEEE Trans. Neural Syst. Rehabil. Eng. **22**(3), 501–510 (2014)

4. F. Clemente, M. D'Alonzo, M. Controzzi, B.B. Edin, C. Cipriani, Non-invasive, temporally discrete feedback of object contact and release improves grasp control of closed-loop myoelectric transradial prostheses. IEEE Trans. Neural Syst. Rehabil. Eng. **24**(12), 1314–1322 (2015)
5. M. Santello, M. Flanders, J.F. Soechting, Postural hand synergies for tool use. J. Neurosci. **18**(23), 10105–10115 (1998)
6. M. Santello, J.F. Soechting, Force synergies for multifingered grasping. Exp. Brain Res. **133**(4), 457–467 (2000)
7. E.J. Weiss, M. Flanders, Muscular and postural synergies of the human hand. J. Neurophysiol. **92**(1), 523–535 (2004)
8. M. Santello, G. Baud-Bovy, H. Jörntell, Neural bases of hand synergies. Front. Comput. Neurosci. **7**, 23 (2013)
9. M. Santello et al., Hand synergies: integration of robotics and neuroscience for understanding the control of biological and artificial hands. Phys Life Rev **17**, 1–23 (2016)
10. R.S. Johansson, J.R. Flanagan, Coding and use of tactile signals from the fingertips in object manipulation tasks. Nat. Rev. Neurosci. **10**(5), 345–359 (2009)
11. J.R. Flanagan, S. King, D.M. Wolpert, R.S. Johansson, Sensorimotor prediction and memory in object manipulation. Can. J. Exp. Psychol. **55**(2), 87–95 (2001)
12. C. Antfolk, M. D'Alonzo, B. Rosen, G.N. Lundborg, F. Sebelius, C. Cipriani, Sensory feedback in upper limb prosthetics. Expert Rev. Med. Devices **10**(1), 45–54 (2013)
13. Q. Fu, M. Santello, Improving fine control of grasping force during hand–object interactions for a soft synergy-inspired myoelectric prosthetic hand. Front. Neurorobot. **11**, 1–15 (2018)
14. M.G. Catalano, G. Grioli, E. Farnioli, A. Serio, C. Piazza, A. Bicchi, Adaptive synergies for the design and control of the Pisa/IIT SoftHand. Int. J. Robot. Res. **33**(5), 768–782 (2014)
15. S.B. Godfrey et al., The Softhand Pro: functional evaluation of a novel, flexible, and robust myoelectric prosthesis. PLoS One **13**(10), 1–20 (2018)
16. S. Casini, M. Morvidoni, M. Bianchi, M.G. Catalano, G. Grioli, A. Bicchi, Design and realization of the CUFF – clenching upper-limb force feedback wearable device for distributed mechano-tactile stimulation of normal and tangential skin forces. IEEE/RSJ Int. Conf. Intell. Robot. Syst., 1186–1193 (2015)
17. Q. Fu, F. Shao, M. Santello, Inter-limb transfer of grasp force perception with closed-loop hand prosthesis. IEEE Trans. Neural Syst. Rehabil. Eng. **27**(5), 927–936 (2019)
18. S. Fani et al., Assessment of myoelectric controller performance and kinematic behavior of a novel soft synergy-inspired robotic hand for prosthetic applications. Front. Neurorobot. **10**, 11 (2016)
19. R.S. Johansson, G. Westling, Roles of glabrous skin receptors and sensorimotor memory in automatic control of precision grip when lifting rougher or more slippery objects. Exp. Brain Res. **56**(3), 550–564 (1984)
20. G. Westling, R.S. Johansson, Factors influencing the force control during precision grip. Exp. Brain Res. **53**(2), 277–284 (1984)
21. A.M. Gordon, G. Westling, K.J. Cole, R.S. Johansson, Memory representations underlying motor commands used during manipulation of common and novel objects. J. Neurophysiol. **69**(6), 1789–1796 (1993)
22. R.S. Johansson, G. Westling, Coordinated isometric muscle commands adequately and erroneously programmed for the weight during lifting task with precision grip. Exp. Brain Res. **71**, 59–71 (1988)
23. R.S. Johansson, K.J. Cole, Sensory-motor coordination during grasping and manipulative actions. Curr. Opin. Neurobiol. **2**(6), 815–823 (1992)
24. A.M. Hadjiosif, M.A. Smith, Flexible control of safety margins for action based on environmental variability. J. Neurosci. **35**(24), 9106–9121 (2015)
25. S.L. Gorniak, V.M. Zatsiorsky, M.L. Latash, Manipulation of a fragile object. Exp. Brain Res. **202**, 413–430 (2010)
26. E.D. Engeberg, S.G. Meek, M.A. Minor, Hybrid force – velocity sliding mode control of a prosthetic hand. IEEE Trans. Biomed. Eng. **55**(5), 1572–1581 (2008)

27. E.D. Engeberg, S.G. Meek, Adaptive sliding mode control for prosthetic hands to simultaneously prevent slip and minimize deformation of grasped objects. IEEE Trans. Mechatron. **18**(1), 376–385 (2013)
28. E.D. Engeberg, M. Frankel, S.G. Meek, Biomimetic grip force compensation based on acceleration of a prosthetic wrist under sliding mode control. IEEE Int. Conf. Robot. Biomimetics, 210–215 (2009)
29. E. Rombokas, C.E. Stepp, C. Chang, M. Malhotra, Y. Matsuoka, Vibrotactile sensory substitution for electromyographic control of object manipulation. IEEE Trans. Biomed. Eng. **60**(8), 2226–2232 (2013)
30. S. Raspopovic et al., Restoring natural sensory feedback in real-time bidirectional hand prostheses. Sci. Transl. Med. **6**(222), 222ra19 (2014)
31. C. Cipriani, F. Zaccone, S. Micera, M.C. Carrozza, On the shared control of an EMG-controlled prosthetic hand: analysis of user-prosthesis interaction. IEEE Trans. Robot. **24**(1), 170–184 (2008)
32. S. Amsuess et al., Context-dependent upper limb prosthesis control for natural and robust use. IEEE Trans. Neural Syst. Rehabil. Eng. **24**(7), 744–753 (2016)
33. N. Jiang, S. Muceli, B. Graimann, D. Farina, Effect of arm position on the prediction of kinematics from EMG in amputees. Med. Biol. Eng. Comput. **51**(1–2), 143–151 (2013)
34. G.K. Patel, J.M. Hahne, C. Castellini, D. Farina, S. Dosen, Context-dependent adaptation improves robustness of myoelectric control for upper-limb prostheses. J. Neural Eng. **14**(5), 056016 (2017)
35. M. Markovic, S. Dosen, D. Popović, B. Graimann, D. Farina, Sensor fusion and computer vision for context-aware control of a multi degree-of-freedom prosthesis. J. Neural Eng. **12**(6), 066022 (2015)
36. M. Markovic, S. Dosen, C. Cipriani, D. Popović, D. Farina, Stereovision and augmented reality for closed-loop control of grasping in hand prostheses. J. Neural Eng. **11**(4), 046001 (2014)
37. C. Castellini et al., Proceedings of the first workshop on peripheral machine interfaces: going beyond traditional surface electromyography. Front. Neurorobot. **8**, 1–17 (2014)
38. L.A. Jones, The Control and Perception of Finger Forces, in *The Human Hand as an Inspiration for Robot Hand Development*, (Springer, Cham, 2014), pp. 99–122
39. S.C. Gandevia, Kinesthesia: Roles for Afferent Signals and Motor Commands, in *Handbook of Physiology: Sec.12. Exercise Regulation and Integration of Multiple Systems*, ed. by L. B. Rowell, J. T. Shepherd, (Oxford University Press, New York, 1996), pp. 128–172
40. M. Haruno, D.M. Wolpert, M. Kawato, MOSAIC model for sensorimotor learning and control. Neural Comput. **13**, 2201–2220 (2001)
41. G. Lafargue, J. Paillard, Y. Lamarre, A. Sirigu, Production and perception of grip force without proprioception: is there a sense of effort in deafferented subjects? Eur. J. Neurosci. **17**(12), 2741–2749 (2003)
42. L.A. Jones, Perceptual constancy and the perceived magnitude of muscle forces. Exp. Brain Res. **151**(2), 197–203 (2003)
43. L.A. Jones, E. Piateski, Contribution of tactile feedback from the hand to the perception of force. Exp. Brain Res. **168**(1–2), 298–302 (2006)
44. S.S. Shergill, P.H. Bays, C.D. Frith, D.M. Wotpert, Two eyes for an eye: the neuroscience of force escalation. Science (80-.) **301**(5630), 187 (2003)
45. R. Ackerley, I. Carlsson, H. Wester, H. Olausson, H. Backlund Wasling, Touch perceptions across skin sites: differences between sensitivity, direction discrimination and pleasantness. Front. Behav. Neurosci. **8**, 1–10 (2014)

Design of a Soft Glove-Based Robotic Hand Exoskeleton with Embedded Synergies

Martin K. Burns and Ramana Vinjamuri

Abstract Paralysis caused by stroke, traumatic brain injury, spinal cord injury, or other injuries and medical conditions can significantly impair the hand function. This impacts an individual's ability to perform manual tasks in activities of daily living (ADL), affecting their independence and psychological well-being. Robotic hand exoskeletons and orthoses have been developed to assist individuals with hand paralysis in conducting ADL; however, existing commercial systems tend to provide basic whole-hand opening and closing motions rather than dexterous manipulation. In this chapter, we document the design of the second prototype of the hand exoskeleton with embedded synergies (HEXOES). This soft glove-based cable-driven robot provides independent actuation of 10 degrees of freedom of the hand using a remote actuator assembly and a lightweight hand component. The hand component, weighing 258 g, actuates the metacarpophalangeal (MCP) and proximal interphalangeal (PIP) joints of each finger and thumb in flexion, with passive extension provided by adjustable springs. Design features were incorporated which aid in donning and doffing and allow individuals with various hand sizes to use the same exoskeleton effectively. Flex sensors placed over the actuated joints along with position and force sensors on the robot's linear actuators enable closed-loop control.

Keywords Wearable robotics · Soft robotics · Hand exoskeleton · Sensor feedback · Kinematic synergies · Assistive robotics · Cable-driven exoskeleton · Hand paralysis · Hand rehabilitation · Object grasping

---

M. K. Burns · R. Vinjamuri (✉)
Stevens Institute of Technology, Hoboken, NJ, USA
e-mail: ramana.vinjamuri@stevens.edu

R. Vinjamuri (ed.), *Advances in Motor Neuroprostheses*,
https://doi.org/10.1007/978-3-030-38740-2_5

1 Introduction

Stroke is one of the most prevalent causes of upper-limb paralysis and impaired function. Estimates from the National Institutes of Health indicate an annual stroke incidence of 795,000 cases, 610,000 of which are first occurrences. This leads to a US stroke prevalence of 6.8 million individuals over 20 years of age [1]. Physical function scores were observed to decrease from 4 months to 16 months poststroke in a Lund, Sweden, study on the quality of life in stroke survivors. After 16 months poststroke, 19% of participants scored in the moderate dependency category of the Barthel Index [2], while 13% were classified with major dependency. The same study found that the mental and emotional quality of life of the participant's informal caregivers (spouses, friends, family) was also correlated with this degree of dependency [3]. This indicates that increasing functional independence in the stroke population yields emotional, social, and financial benefits for the individual as well as those around them. In the context of this chapter, functional independence through hand grasping is discussed.

Increasing functional independence in hand grasping can be approached through rehabilitation, through daily or periodic exercises, or by using assistive devices. The latter has undergone extensive development in recent years in the field of assistive and rehabilitative wearable robotics. An overview of robotic rehabilitation devices is presented in [4], covering advances using hand trainers, arm exoskeletons, and BCI-based systems in clinical studies. Among these works is the rehabilitative HWARD exoskeleton, which demonstrated that increased robotic hand therapy led to improvement in motor function scores [5], and the Reha-Digit, which demonstrated minor improvements in functional scores when doing additional exercises with a finger trainer in subacute stroke patients [6]. Passive exoskeleton devices such as the HandSOME have also shown significant functional recovery in patients with stroke during at-home therapy [7]. Additional studies compared brain-computer interface (BCI)-driven devices to robot-only and standard therapy, with the hypothesis that integrating the patient's movement intent would lead to improved outcomes over purely robotic training. These works found significant improvements in functional recovery with a haptic knob [8], hand exoskeletons [9–11], and arm orthoses [12]. A clinical trial studying BCI training with a wrist orthosis yielded no improvement in hand motor function; however, effective control of the system using a BCI was achieved [13]. BCI/FES systems are also being studied, with favorable preliminary results observed in arm and hand-based exoskeleton rehabilitation [14].

In addition to rehabilitative applications, wearable robots can be used to restore the hand function through assistance during activities of daily living (ADL). Rehabilitation devices do not need to be carried or worn for an extended period, so they tend to be rigid, immobile systems designed to provide force and control to the user's hand. Alternatively, assistive devices meant to be worn throughout the day to assist in ADL grasping must, above all, be compact and lightweight to avoid fatiguing the user [15, 16]. A large amount of research has investigated lightweight mechanisms, actuators, materials, and structures to design functional wearable robots which do not over-encumber the wearer. Lightweight pneumatic actuators which are cast from silicon or some other polymer were investigated by [17, 18]. These devices bend when pressurized based on the structure's varying stiffness. Further, work integrated fiber into the polymer body to control twisting and flexing motions [19]. These actuators were integrated into the Wyss Institute's Soft Robotic Glove [20, 21] and were used in a separate device [22]. Another group made improvements to the traditional McKibben actuators drastically reduced the required operating pressures [23]. These improved actuators were integrated into a lightweight wrist orthosis [24]. The behavior of cable-driven soft robotic gloves has been studied [25, 26], leading to several devices in the recent research. These include the Exo-Glove Poly II [27], the SEM Glove, now available as the Carbonhand [28], and the Flexo-glove [29], the mano [30], among others. A hybrid device called the AirExGlove was also produced, which used both a cable system on the palmar side and pneumatic actuators on the dorsal side to achieve antagonistic actuation [31].

These soft devices have managed to actuate low-DoF hand opening and closing motions with lightweight structures. However, they are not able to provide high-DoF actuation which would be needed for dexterous object manipulations. An ideal system would match the functional dexterity of the hand; however, this would lead to a new problem of control. The traditional control methods operate on 6 DoF systems, or 7 DoF with redundancy; however, adding more DoFs leads to an exponentially more complicated control problem. Furthermore, an individual operating a prosthetic or exoskeleton will only be able to consciously operate one to three DoF. In order to address this problem, research in grasp biomechanics has determined that dimensionality reduction can be leveraged in the form of synergies.

Synergies, or "a collection of relatively independent degrees of freedom that behave as a single functional unit" [32], are believed to underlie the neural control mechanisms used during the normal hand grasping. Early animal model experiments revealed that stimulation of individual spinal cord neurons elicited coupled activations of limb muscles [33, 34] and that the resulting motion patterns were similar to the synergies extracted directly from the observed kinematics [35]. The central nervous system, therefore, is believed to control these combined movement patterns to achieve dexterous grasping rather than individually coordinate the joints of the hand [36]. Since then, many studies have derived postural synergies from human grasp biomechanics [37–41]. Postural synergies encode relative movement patterns across the joints which are fixed in time; however, spatiotemporal synergies have also been found which accurately replicate hand grasping [42–44] and bilateral arm movements [45]. The most recent work on hand synergies has translated them from biomechanical analysis techniques to practical robotic systems. Postural synergies have been incorporated into control systems for robotic hands such as the DLR Hand [46, 47] and the UB Hand IV [48], as well as human-machine interfaces for robotic arms [49] and EMG-driven prosthetic hands [50]. Whereas most works are on prosthetic devices for amputees, some wearable exoskeletons are currently in development which aim to control a paralyzed hand using synergies [51, 52]. These devices hold promise for the large population of individuals with stroke; however, a reliable test bed for synergy-based assistive and rehabilitative controls in a wearable system must be developed.

There are several requirements we used to develop a wearable synergy-based testing system. These criteria are based on experience both within the Sensorimotor Control Laboratory, and from qualities previously identified in the literature.

1. Wearable: The worn part of the synergy exoskeleton must be lightweight and easy for a user or assistant to don and doff. This is essential to make the system practical for laboratory studies with human subjects, as well as for future devices meant for at-home use.
2. Sizeable: The device should be designed to fit as many individuals as possible. This would make a single device usable in a larger portion of affected individuals.
3. Safe: The worn device must be safe to operate for the user. This includes avoiding designs which may injure the user's hand by hard edges or pinch points, as well as including safety stops in hardware or software.
4. High DoF: The synergies that the device needs to actuate represent high-DoF whole-hand motions, so the device needs to have the capability to independently actuate a high number of DoF across the hand.
5. Powerful: The device must be powerful enough to actuate the hands of healthy individuals and individuals with paralysis, including those with a spastic hand reflex.
6. Sensor feedback: The system must have adequate sensor feedback for closed-loop control, to ensure the hand is actuated along the synergies being studied.
7. Configurable synergies: The synergies that the device uses must be configurable, so that movement profiles from any source (intra- or intersubject synergies, mathematically optimized synergies, spatial or spatiotemporal synergies, etc.) can be used for control.

A device which meets these criteria would be an adequate platform on which synergy-based control methods can be developed and tested.

In this chapter, we describe the hand exoskeleton with embedded synergies (HEXOES). The HEXOES is a 10-DoF cable-driven soft exoskeleton with a remote actuator unit. The device provides active flexion and passive extension of the MCP and PIP/IP joints of each finger and thumb. The purpose of this prototype is to provide enough actuated degrees of freedom (DoF) to test synergy-based assistive and rehabilitative control systems for individuals with hand paralysis. This design makes several improvements over the preceding iteration [51] in weight, usability, and sensor accuracy.

2 Methods

The HEXOES is built from two systems: an actuator assembly and a soft glove-based hand component (Fig. 1a). The actuator assembly and hand component are connected using a bundle of Bowden tubes and a ribbon cable. The prototype is capable of independently actuating flexion of the metacarpophalangeal (MCP) and interphalangeal/proximal interphalangeal (IP/PIP) joints of the thumb and fingers,

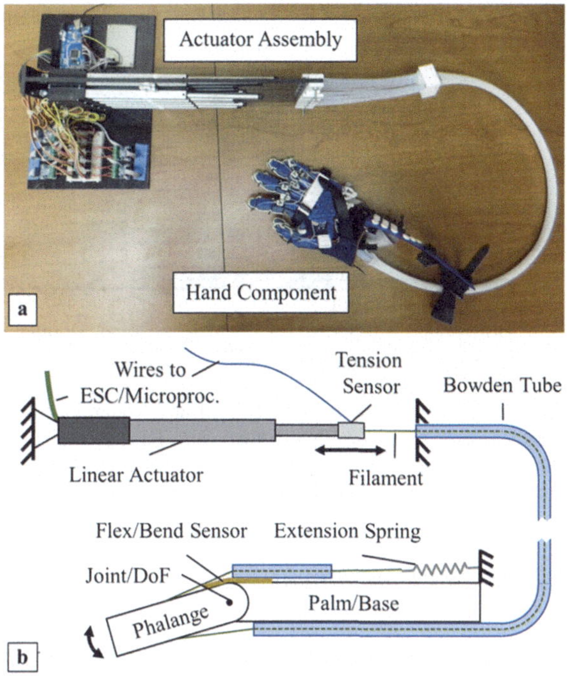

Fig. 1 (**a**) HEXOES actuator assembly and hand component. (**b**) Actuation diagram for a 1 DoF MCP joint of the exoskeleton. A linear actuator and tension sensor applies force to a filament tendon, which anchors to the phalange through a Bowden tube to provide actuated flexion. An adjustable spring provides passive extension, and a flex sensor detects bending at the joint

with extension passively actuated by adjustable springs. The ten actuated DoFs are monitored using several sensors, allowing closed-loop control systems in the future. The design of the actuator assembly will first be described, followed by the hand component.

2.1 Actuator Assembly

The actuator assembly, shown in Fig. 2, consists of linear actuators and drive electronics mounted to ABS sheets and weighs 2.2 kg. External power for the motors and speed controllers is supplied from a 12 V benchtop DC power source limited to 3A of current. The linear actuators have a stroke length of 140 mm, peak force of 100 N, and a no-load maximum speed of 20 mm/s. Each actuator provides positional feedback through a built-in potentiometer, which returns a voltage ranging from 0.5 to 4.5 V. A miniature force sensor (SMD Sensors, Wallingford CT, USA) is mounted on the end of each actuator which provides tension measurement for each actuated DoF up to 44 N. High-strength braided filament tendons attached to each sensor enter PTFE filament guides, which converge to a bundle of Bowden tubes wrapped in a mesh sleeve. The tube bundle also houses a ribbon cable which carries signals from the hand component's joint angle sensors. This cable terminates at a separate Arduino meant for analog inputs from the hand's flex sensors.

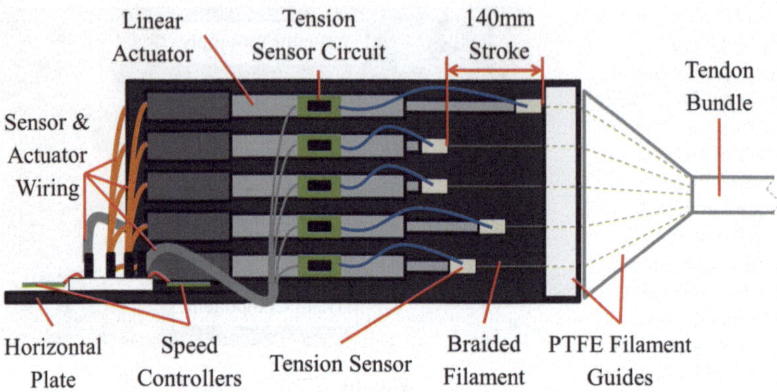

Fig. 2 HEXOES actuator assembly, with major components labeled. Ten linear actuators with tension sensors are mounted in vertical pairs. Each sensor is attached to a braided filament which routes into a tendon bundle containing ten Bowden tubes. Tension sensor circuits, electronic speed controllers, and microcontrollers are also included in this assembly

2.2 Hand Component

Since the hand component is in direct contact with the user, its design must be carefully considered to ensure it is a usable, wearable device. The physical design of the system must be of minimal size and weight so as not to encumber the user yet must be robust enough to manipulate the hand. The dimensions of the exoskeleton and compliance of the chosen materials should fit a wide range of hand sizes. The hand structure must include anchor points and tendon routing that results in the actuation of the MCP and PIP joints of the fingers and thumb for a total of 10 actuated degrees of freedom. Each of these joints should also have a sensor to provide joint angle feedback. Since the actuator assembly can only provide flexion forces through tension on the braided filament, the exoskeleton must incorporate passive extension. Finally, the hand must be easy to don and doff with the help of an experimenter or clinician.

The palmar and dorsal vies of the hand component are shown in Fig. 3a and b, respectively. A thin nitrile-coated work glove serves as the base for the assembly, with several components stitched onto the material. Flexible 3D printed thermoplastic polyurethane (TPU) parts are designed for the fingers and forearm brace to allow compliance for hand sizing. Inter-digit TPU spacers provide abduction to the fingers to prevent collisions during grasping. A solid palm brace printed from polylactic acid (PLA) and mounted onto a bendable metal wrist band provides wrist support and slight wrist dorsiflexion. Nylon cloth pockets are stitched over the MCP and PIP/DIP joints of the hand which house flex sensors to allow joint angle estimation. The palm support also houses the distal ends of the Bowden tube filament guides, as well as an adjustable pivoting thumb opposition hook. The forearm brace features Velcro straps, an electrical connector for the sensors, and buttons used for adjusting the passive extension springs. The total mass of the assembly is 258 g, which compares favorably to other hand exoskeleton systems in the literature.

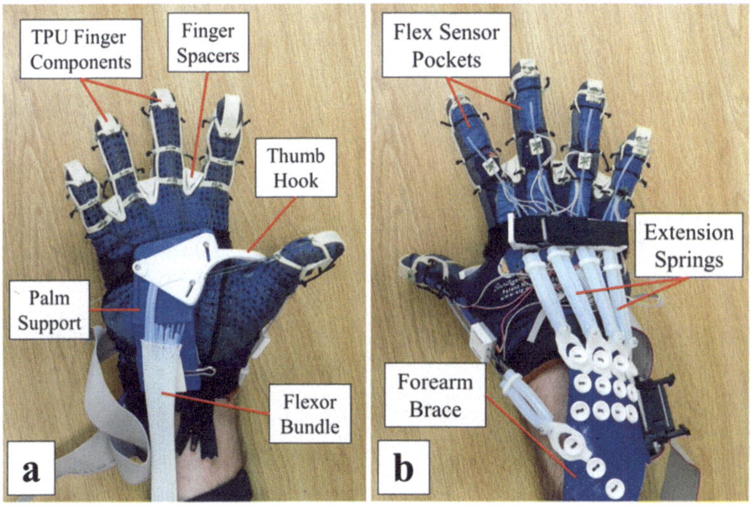

Fig. 3 Hand component of the HEXOES with major parts labeled. (**a**) Dorsal view of the hand, showing the palm support, thumb opposition hook, tendon bundle insertion, flexion tendons, TPU finger components, and finger spacers. (**b**) Dorsal side of the HEXOES, showing the adjustable extension springs, sensor pockets, and Velcro straps

Each finger component is a single 3D printed part made from TPU and serves several roles on the exoskeleton. The part is designed in SolidWorks in an unwrapped state (Fig. 4a), 3D printed flat, and wrapped into its final state (Fig. 4b) before being stitched onto the work glove. This flat-print method has several benefits to a print-as-is approach, in which the part is printed in the exact form it will be mounted in. Complex geometries in the final part would be difficult to print directly but can be easily printed in an unwrapped 2D profile. This enables the use of consumer-level FDM printers to produce otherwise prohibitively difficult designs. The 2D print orientation also ensures the layer boundaries are along the maximal cross section of the part, reducing or eliminating the risk of part failure. Each finger component is designed according to measurements taken from a virtual hand model (CADHuman, South Carolina, USA) scaled from a fifth percentile female to a 95th percentile male. The distances between the fingertip and the MCP, PIP, and DIP joints are measured for each size. The measurements are then imposed relative to a virtual fingertip datum to determine bounds for the locations of the joint centers of rotation, indicated by the shaded region in Fig. 4a. A series of tendon anchor points, stitch holes, and elastic holes are positioned relative to the joint axis regions. The fingertip includes stitch holes which mate with legs designed to wrap from the palmar to dorsal side of the finger, creating a cup around the user's distal phalange. The elastic on the dorsal side of the hand provides radial compression against the user's fingers, counteracting the glove deformation characteristic of soft tendon-driven exoskeleton gloves. The elastic cord and deformable plastic also provide a high degree of compliance in the fingers, allowing a range of finger widths to wear

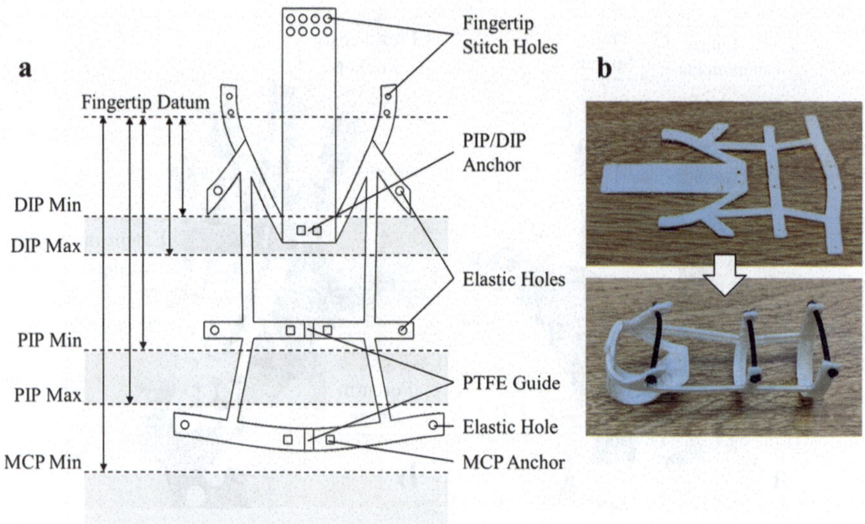

Fig. 4 (**a**) Dimensions and features of the unwrapped CAD model of the finger components. (**b**) The flat, printed finger component is wrapped before attaching to the exoskeleton

the exoskeleton. The components also include channels to mount PTFE tendon guides on the palmar side of the finger. The flat printed finger components are wrapped into their functional state (Fig. 4b), and stitched onto the work glove.

Several features on the glove are meant to ease the process of donning and doffing the exoskeleton. The palm of the exoskeleton can be opened using a pair of zippers and a mechanical clasp on the palm support, as shown in Fig. 5. The clasp uses a peg with a dovetail profile, mounted to the bendable metal wrist band, and a matching groove on the palm support piece. With these features disengaged, the palm of the exoskeleton can be opened, giving an easy access to the glove's fingers for a paralyzed individual or an assistant to put the glove on the user. Once the user's hand is in the exoskeleton, the zippers can be closed, and the palm support can be clasped in place. The palm support slides onto the peg, and a channel through both pieces allows a cotter pin to be inserted to lock the assembly together. The thumb opposition hook can then be positioned to guide the thumb into a comfortable grasping position.

2.3 *Electrical Design*

The HEXOES is composed of ten individual actuators and thirty sensors for the ten independent DoF. The architecture, shown in Fig. 6, is centered on an STM32F446 Nucleo-64 microcontroller which interfaces with a personal computer (PC) over USB (Universal Serial Bus). The microcontroller operates an Arm Cortex-M4 core

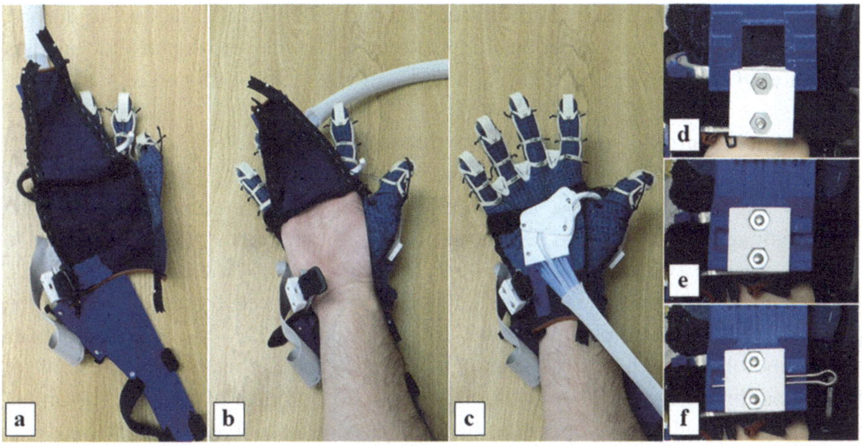

Fig. 5 Clasp and zipper system for the HEXOES donning and doffing. (**a**) The zippers and clasp are open, allowing full access to the palm and fingers. (**b**) The hand is placed in the glove, and (**c**) the zippers are closed around the hand. (**d**) The wrist band is bent into place, and (**e**) the groove of the palm support is slid onto the dovetail peg on the wrist band. (**f**) A cotter pin is inserted to lock the mechanism

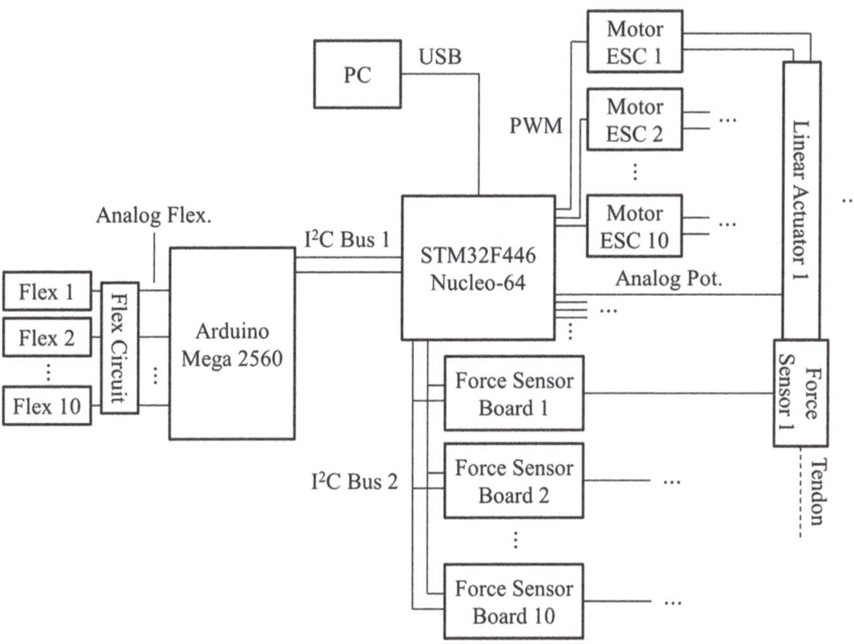

Fig. 6 Electrical architecture of the HEXOES. An STM32F446 Nucleo board interfaces with motor electronic speed controllers (ESCs) using PWM, with the force sensor boards over I^2C, and with the flex sensors via an Arduino Mega 2560 over a separate I^2C bus

at 180 MHz with 512 KB of flash memory and 128 KB of ram. In the current stage the PC implements all control algorithms while the microcontroller handles hardware I/O; however, the STM32 board can run embedded controls in future versions.

The microcontroller interfaces with three different types of sensors. The flex sensors in the hand component are connected to 33 KΩ voltage dividers operating at a 4.5 V excitation voltage. An Arduino Mega 2560 microcontroller reads the analog voltages, and the STM32 controller reads these signals over a 100 kHz I^2C bus[1]. A separate 100 kHz bus is used to interface with the ten force sensor boards, which return signals that are converted to factory-calibrated force values. The ten potentiometers are read using the analog pins on the STM32. The STM32 controls the ten motors through five two-channel electronic speed controllers (ESCs) using PWM and direction pins. Data is filtered using a 60 Hz notch filter and a 10-sample moving average filter. The operating frequency of the overall system is 500 Hz.

This architecture can function either as a low-level input/output system or as a fully embedded system. In the former case, all user inputs, signal processing, synergy extraction, control algorithms, and/or models would be executed on a computer. The computer would read raw sensor values and write individual motor commands to the exoskeleton over the USB drive. The microcontroller would then handle all internal communication and motor driver logic. In the latter configuration, the biosignal processing, extraction of synergy recruitments, and data recording would be handled by a PC. The computer would communicate synergy recruitment commands to the exoskeleton over USB. The exoskeleton would accept these synergy recruitments and execute the closed-loop controls internally to actuate the hand. The microcontroller can also be configured for wireless operation or can be integrated into a finalized system including user inputs.

3 Discussion

The exoskeleton presented here was designed to incorporate several improvements over the first HEXOES [51] while making several compromises to create a functional synergy-based test system. The primary concerns were to minimize the weight of the hand component and ease donning and doffing of the system while maintaining a high number of independently actuated DoF.

There is no definitive requirement for the maximum mass that a hand exoskeleton can have and still be usable; however, Polygerinos et al. [21] have outlined a maximum wearable actuator assembly mass of 3 kg and a hand component mass of 500 g, while another project outlined a mass requirement for a pediatric wrist/thumb exoskeleton of approximately 70% the mass of a child's arm, or 450 g [53]. These

[1]Level converters were used to connect the 3.3 V STM32 bus to the 5 V Arduino and force sensor boards.

estimates were based on feedback from clinical collaborators associated with the respective projects. Using an estimate of 426 g for the mass of an adult male hand [54] and assuming 70% mass allowance yield a maximum hand component mass of 300 g. The hand component of the HEXOES satisfies this requirement with a total mass of 258 g and a total actuator assembly mass of 2.2 kg.

Donning and doffing the hand exoskeleton is the other major issue considered in the HEXOES. Spasticity, which is a continuous involuntary contraction of one or more muscles, is a common phenomenon in individuals with stroke, occurring in anywhere from 18% to 42.6% of stroke survivors [55–57], with some estimates reaching 65% [58]. No reliable data is available for the prevalence of spasticity of the hand; however, one study showed that the wrist is affected in 66% of those with spasticity [57]. This condition can manifest as a constant hand closing reflex, making it difficult to put a glove exoskeleton on the user. Relocating the actuators to a remote assembly alleviated the bulk that impeded donning, while several design decisions in the hand component further reduced the effort required to put on the exoskeleton.

Based on our objective of producing a synergy test bed, we designed a system with a large number of actuated DoF which can be controlled using any synergies. This requirement implies that software-level synergy control is the most appropriate, since these synergies can be changed readily. However, several groups have developed systems with mechanically implemented synergies. The most direct way to implement synergies is using a pulley system, which was first accomplished using a pair of shafts with pulleys with radii corresponding to postural synergy weights [59]. This was later replicated in another two-synergy hand using pulleys and was controlled using myoelectric signals in able-bodied subjects and an amputee [60]. Recently, a soft hand exoskeleton actuated with one synergy was designed [52] and validated [61] which used a similar pulley system to flex and extend the thumb, index, and middle finger. These groups implemented postural synergies using pulleys; however, another group designed a mechanical hand actuated using two spatiotemporal synergies. This was done by replacing the constant-radii pulleys for each joint with varying-radii cams, resulting in a series of tendon retractions which vary over the time span of the synergy [62]. The concept of pulley-driven synergy hands was refined into the PISA/IIT Softhand, a completely self-contained prosthetic hand which used one synergy [63] and two synergies [64] to drive the 19-DoF compliant hand. Although pulley-based mechanisms are the most common, alternative implementations of postural synergies have also been prototyped. A 19-DoF prosthetic hand was actuated with two synergies using a series of stacked planetary gears [65]. The same group later designed a 19-DoF hand which was actuated using two synergies with a composed continuum mechanism [66].

These designs accomplished complex hand movements using only one or two actuators, significantly reducing the total mass of the devices; however a major drawback is that their movement profiles are not easily configurable. Each project has extracted synergies from different grasp datasets; however, there is currently no evaluation on what synergies are best for ADL grasping and manipulations. A new device would need to be fabricated to actuate according to a different set of synergy profiles, which is a significant hurdle to evaluating the effectiveness of synergies in everyday assistance. For this reason, we accept the additional weight in the actuator assembly from the ten motors since it does not affect the mass of the hand component. This compromise will allow us to test synergy-based systems which are configurable on a subject-by-subject basis.

An important distinction to be made for an exoskeleton or other robotic wearable device is whether it is meant to be an assistive or rehabilitative system. These two classes of device have mutually exclusive design requirements. An assistive device requires minimal weight and maximum mobility and must provide enough motion to restore grasping ability. This can be accomplished by current exoskeletons with as few as 1 DoF. A rehabilitation system, however, would require a mechanically robust system that can finely control the motion of the hand, without as much need for mobility. The Armeopower/Manovopower, Hand of Hope, and BRAVO exoskeletons represent the rehabilitation side of this trade-off. The HEXOES aims to evaluate control systems for both assistive and rehabilitative applications. The lightweight hand module allows natural grasp studies to be performed so that assistive systems can be tested in a controlled lab environment; however, the 10 DoF actuator assembly, weighing 2.2 kg, is not mobile enough to be used as a daily assistive system unless mounted on a wheelchair. This prototype does, alternatively, provide sufficient DoF and actuator speed and strength to implement rehabilitation systems. However, the soft hand design may be too compliant to conduct effective rehabilitation exercises independent of researchers or physical therapists. By accepting these compromises, we aim to make a system that can evaluate the feasibility of assistive and rehabilitative control systems. Subsequent hand exoskeleton designs can be targeted to a specific application once the underlying control theory has been validated.

Future work with the HEXOES includes immediate progress on closed-loop control integration and long-term progress on scientific exploration of synergies and application-specific designs. In the short term, closed-loop synergy controls such as those presented in [67] will be integrated into the HEXOES system. This will require robust joint angle feedback and model-based control. We will continue our ongoing work on extracting synergy controls from user inputs, including common interfaces such as joysticks, buttons, and remote controls as well as noninvasive biosignals such as electromyography (EMG) and electroencephalography (EEG). Invasive biosignals such as electrocorticography (ECoG) or neural/spinal implants could also be used to extract synergy commands. The HEXOES can also supply sensor feedback to the user using its numerous sensors, closing the loop between the user's proprioception and the device's movement commands. Long-term progress will include experiments on assistive and rehabilitative control strategies based on synergies, and descendent exoskeleton/prosthetic designs can be informed by these experiments with the HEXOES test bed. This can include custom-sized or population-sized hand components, as well as designs refined specifically for assistance versus rehabilitation. Mechanically implemented synergies based on experimental results can be incorporated to reduce cost and weight.

4 Conclusion

This chapter details the mechanical and electrical design of the HEXOES, the Hand Exoskeleton with Embedded Synergies. This device aims to serve as a test bed for synergy-based assistive control systems and rehabilitative training protocols. It provides actuation and sensor feedback for the ten most commonly studied DoFs of the hand. The soft structure of the hand component is comfortable to wear and includes a zipper and clasp system that makes the system easy to don and doff. The soft structure also results in a low-weight system which does not sacrifice device actuation. The microcontroller can be run as an I/O device operated by a PC for rapid experimental research or as an embedded synergy-driven system. The HEXOES, therefore, is an adequate test system for wearable synergy-based systems. Future developments for this system include closed-loop control integration, experiments with synergy controls, and future prototypes for specific applications.

References

1. V.Y. Ma, L. Chan, K.J. Carruthers, The incidence, prevalence, costs and impact on disability of common conditions requiring rehabilitation in the US: stroke, spinal cord injury, traumatic brain injury, multiple sclerosis, osteoarthritis, rheumatoid arthritis, limb loss, and back pain. Arch. Phys. Med. Rehabil. **95**, 986–995 (2014). https://doi.org/10.1016/j.apmr.2013.10.032.The
2. F.I. Mahoney, D.W. Barthel, Functional evaluation: the Barthel index. Md. State Med. J. **14**, 56–61 (1965)
3. A.-C. Jonsson, I. Lindgren, B. Hallstrom, et al., Determinants of quality of life in stroke survivors and their informal caregivers. Stroke **36**, 803–808 (2005). https://doi.org/10.1161/01.STR.0000160873.32791.20
4. A.A. Frolov, I.B. Kozlovskaya, E.V. Biryukova, P.D. Bobrov, Use of robotic devices in post-stroke rehabilitation. Neurosci. Behav. Physiol. **48**, 1053–1066 (2018). https://doi.org/10.1007/s11055-018-0668-3
5. C.D. Takahashi, L. Der-Yeghiaian, V. Le, et al., Robot-based hand motor therapy after stroke. Brain **131**, 425–437 (2008). https://doi.org/10.1093/brain/awm311
6. S. Hesse, H. Kuhlmann, J. Wilk, et al., A new electromechanical trainer for sensorimotor rehabilitation of paralysed fingers: a case series in chronic and acute stroke patients. J. Neuroeng. Rehabil. **5**, 2–7 (2008). https://doi.org/10.1186/1743-0003-5-21
7. J. Chen, D. Nichols, E.B. Brokaw, P.S. Lum, Home-based therapy after stroke using the hand spring operated movement enhancer (HandSOME). IEEE Trans. Neural Syst. Rehabil. Eng. **25**, 2305–2312 (2017). https://doi.org/10.1109/TNSRE.2017.2695379
8. K.K. Ang, C. Guan, K.S. Phua, et al., Brain-computer interface-based robotic end effector system for wrist and hand rehabilitation: results of a three-armed randomized controlled trial for chronic stroke. Front. Neuroeng. **7**, 1–9 (2014). https://doi.org/10.3389/fneng.2014.00030
9. A.A. Frolov, O. Mokienko, R. Lyukmanov, et al., Post-stroke rehabilitation training with a motor-imagery-based brain-computer interface (BCI)-controlled hand exoskeleton: a randomized controlled multicenter trial. Front. Neurosci. **11**, 400 (2017). https://doi.org/10.3389/fnins.2017.00400
10. G.E. Ivanova, Y.V. Bushkova, A.Y. Suvorov, et al., Use of a BCI-exoskeleton simulator with multichannel biofeedback in a multidisciplinary rehabilitation program in poststroke patients. Neurosci. Behav. Physiol. **48**, 1100–1105 (2018). https://doi.org/10.1007/s11055-018-0673-6

11. S.V. Kotov, L.G. Turbina, A.A. Kondur, et al., Recovery dynamics in patients with poststroke motor disorders after multiple courses of neurorehabilitation using an exoskeleton controlled by a brain–computer interface. Neurosci. Behav. Physiol. **48**, 1088–1092 (2018). https://doi.org/10.1007/s11055-018-0671-8
12. A. Ramos-Murguialday, D. Broetz, M. Rea, et al., Brain-machine-interface in chronic stroke rehabilitation: a controlled study. Ann. Neurol. **74**, 100–108 (2013). https://doi.org/10.1002/ana.23879
13. E. Buch, C. Weber, L.G. Cohen, et al., Think to move: A neuromagnetic brain-computer interface (BCI) system for chronic stroke. Stroke **39**, 910–917 (2008). https://doi.org/10.1161/STROKEAHA.107.505313
14. M.S. Poboroniuc, D.C. Irimia, FES&BCI based rehabilitation engineered equipment: clinical tests and perspectives. In *2017 E-Health Bioeng Conf EHB 2017* (2017), pp. 77–80. https://doi.org/10.1109/EHB.2017.7995365
15. S. Ates, C.J.W. Haarman, A.H.A. Stienen, SCRIPT passive orthosis: design of interactive hand and wrist exoskeleton for rehabilitation at home after stroke. Auton. Robots **41**, 711–723 (2017). https://doi.org/10.1007/s10514-016-9589-6
16. B.W. Gasser, D.A. Bennett, C.M. Durrough, M. Goldfarb, Design and preliminary assessment of Vanderbilt hand exoskeleton. IEEE Int. Conf. Rehabil. Robot. **2017**, 1537–1542 (2017). https://doi.org/10.1109/ICORR.2017.8009466
17. H.K. Yap, J.H. Lim, F. Nasrallah, et al., A soft exoskeleton for hand assistive and rehabilitation application using pneumatic actuators with variable stiffness. Proc. IEEE Int. Conf. Robot. Autom. **2015**, 4967–4972 (2015). https://doi.org/10.1109/ICRA.2015.7139889
18. H.K. Yap, J.H. Lim, F. Nasrallah, C.H. Yeow, Design and preliminary feasibility study of a Soft Robotic Glove for hand function assistance in Stroke Survivors. Front. Neurosci. **11**, 1–14 (2017). https://doi.org/10.3389/fnins.2018.00323
19. P. Polygerinos, Z. Wang, J.T.B. Overvelde, et al., Modeling of soft fiber-reinforced bending actuators. IEEE Trans. Robot. **31**, 778–789 (2015b). https://doi.org/10.1109/TRO.2015.2428504
20. L. Cappello, J.T. Meyer, K.C. Galloway, et al., Assisting hand function after spinal cord injury with a fabric-based soft robotic glove. J. Neuroeng. Rehabil. **15**, 1–10 (2018). https://doi.org/10.1186/s12984-018-0391-x
21. P. Polygerinos, Z. Wang, K.C. Galloway, et al., Soft robotic glove for combined assistance and at-home rehabilitation. Robot. Auton. Syst. **73**, 135–143 (2015a). https://doi.org/10.1016/j.robot.2014.08.014
22. Y. Jiang, D. Chen, J. Que, et al., Soft robotic glove for hand rehabilitation based on a novel fabrication method. In *2017 IEEE International Conference on Robotics and Biomimetics, ROBIO 2017* (2017), pp 1–6
23. K. Ogawa, C. Thakur, T. Ikeda, et al., Development of a pneumatic artificial muscle driven by low pressure and its application to the unplugged powered suit. Adv. Robot. **31**, 1135–1143 (2017). https://doi.org/10.1080/01691864.2017.1392345
24. T. Goto, Y. Kurita, S. Das, K. Kunze, Artificial motion guidance: an intuitive device based on Pneumatic Gel Muscle (PGM). In *UIST 2018 Adjun – Adjun Publ 31st Annu ACM Symp User Interface Softw Technol* (2018), pp. 182–184. https://doi.org/10.1145/3266037.3271644
25. H.K. In, K.J. Cho, Evaluation of the antagonistic tendon driven system for SNU Exo-Glove. In *2012 9th Int Conf Ubiquitous Robot Ambient Intell URAI 2012* (2012), pp. 507–509. https://doi.org/10.1109/URAI.2012.6463054
26. H.K. In, K.J. Cho, K.R. Kim, B.S. Lee, Jointless structure and under-actuation mechanism for compact hand exoskeleton. IEEE Int. Conf. Rehabil. Robot., 1–6 (2011, 2011). https://doi.org/10.1109/ICORR.2011.5975394
27. B.B. Kang, H. Choi, H. Lee, K.-J. Cho, Exo-glove poly II: a polymer-based soft wearable robot for the hand with a tendon-driven actuation system. Soft Robot **6**, 214–227 (2019). https://doi.org/10.1089/soro.2018.0006

28. M. Nilsson, J. Ingvast, J. Wikander, H. Von Holst, The Soft Extra Muscle system for improving the grasping capability in neurological rehabilitation. In *2012 IEEE-EMBS Conference on Biomedical Engineering and Sciences, IECBES 2012* (2012), pp 412–417
29. A. Mohammadi, J. Lavranos, P. Choong, D. Oetomo, Flexo-glove: a 3D printed soft exoskeleton robotic glove for impaired hand rehabilitation and assistance. Proc Annu Int Conf IEEE Eng Med Biol Soc EMBS **2018**, 2120–2123 (2018). https://doi.org/10.1109/EMBC.2018.8512617
30. L. Randazzo, I. Iturrate, S. Perdikis, J.D.R. Millán, Mano: a wearable hand exoskeleton for activities of daily living and neurorehabilitation. IEEE Robot. Autom. Lett. **3**, 500–507 (2018). https://doi.org/10.1109/LRA.2017.2771329
31. A. Stilli, A. Cremoni, M. Bianchi, et al., AirExGlove-A novel pneumatic exoskeleton glove for adaptive hand rehabilitation in post-stroke patients. In *2018 IEEE Int Conf Soft Robot RoboSoft 2018* (2018), pp. 579–584. https://doi.org/10.1109/ROBOSOFT.2018.8405388
32. M.T. Turvey, Action and perception at the level of synergies. Hum. Mov. Sci. **26**, 657–697 (2007). https://doi.org/10.1016/j.humov.2007.04.002
33. T. Drew, J. Kalaska, N. Krouchev, Muscle synergies during locomotion in the cat: a model for motor cortex control. J. Physiol. **586**, 1239–1245 (2008). https://doi.org/10.1113/jphysiol.2007.146605
34. P. Saltiel, K. Wyler-Duda, A. D'Avella, et al., Muscle synergies encoded within the spinal cord: evidence from focal intraspinal NMDA iontophoresis in the frog. J. Neurophysiol. **85**, 605–619 (2001). https://doi.org/10.1152/jn.2001.85.2.605
35. N. Krouchev, J.F. Kalaska, T. Drew, Sequential activation of muscle synergies during locomotion in the intact cat as revealed by cluster analysis and direct decomposition. J. Neurophysiol. **96**, 1991–2010 (2006). https://doi.org/10.1152/jn.00241.2006
36. M. Santello, G. Baud-Bovy, H. Jörntell, Neural bases of hand synergies. Front. Comput. Neurosci. **7**, 1–15 (2013). https://doi.org/10.3389/fncom.2013.00023
37. C.R. Mason, J.E. Gomez, T.J. Ebner, Hand synergies during reach-to-grasp. J. Neurophysiol. **86**, 2896–2910 (2001). https://doi.org/10.1080/00222895.1993.9942048
38. A. Sahbani, A. Roby-Brami, W. Bachta, et al., Analysis of hand synergies in healthy subjects during bimanual manipulation of various objects. J. Neuroeng. Rehabil. **11**, 113 (2014). https://doi.org/10.1186/1743-0003-11-113
39. M. Santello, M. Flanders, J.F. Soechting, Patterns of hand motion during grasping and the influence of sensory guidance. J. Neurosci. **22**, 1426–1435 (2002)
40. E. Todorov, Z. Ghahramani, Analysis of the synergies underlying complex hand manipulation. 26th Annu. Int. Conf. IEEE Eng. Med. Biol. Soc. **2**, 4637–4640 (2005). https://doi.org/10.1109/iembs.2004.1404285
41. R. Vinjamuri, M. Sun, C.-C. Chang, et al., Dimensionality reduction in control and coordination of the human hand. IEEE Trans. Biomed. Eng. **57**, 284–295 (2010). https://doi.org/10.1109/TBME.2009.2032532
42. V. Patel, M. Burns, R. Vinjamuri, Effect of visual and tactile feedback on kinematic synergies in the grasping hand. Med. Biol. Eng. Comput. **54**, 1217–1227 (2016). https://doi.org/10.1007/s11517-015-1424-2
43. D. Pei, V. Patel, M. Burns, et al., Neural decoding of synergy-based hand movements using electroencephalography. IEEE Access **7**, 1–1 (2019). https://doi.org/10.1109/access.2019.2895566
44. R. Vinjamuri, D.J. Weber, Z.H. Mao, et al., Toward synergy-based brain-machine interfaces. IEEE Trans. Inf. Technol. Biomed. **15**, 726–736 (2011). https://doi.org/10.1109/TITB.2011.2160272
45. M.K. Burns, V. Patel, I. Florescu, et al., Low-dimensional synergistic representation of bilateral reaching movements. Front. Bioeng. Biotechnol. **5**, 2 (2017a). https://doi.org/10.3389/fbioe.2017.00002
46. G. Salvietti, T. Wimboeck, D. Prattichizzo, A static intrinsically passive controller to enhance grasp stability of object-based mapping between human and robotic hands, in *2013 IEEE/RSJ international conference on intelligent robots and systems (IROS)*, ed. by N. Amato, (IEEE, New York, NY, 2013), pp. 2460–2465

47. T. Wimböck, B. Jahn, G. Hirzinger, Synergy level impedance control for multifingered hands. IEEE Int. Conf. Intell. Robot. Syst. **2011**, 973–979 (2011). https://doi.org/10.1109/IROS.2011.6048163
48. F. Ficuciello, G. Palli, C. Melchiorri, B. Siciliano, Postural synergies of the UB Hand IV for human-like grasping. Robot. Auton. Syst. **62**, 515–527 (2014). https://doi.org/10.1016/j.robot.2013.12.008
49. P.K. Artemiadis, K.K. Kyriakopoulos, EMG-based control of a robot arm using low-dimensional embeddings. IEEE Trans. Robot. **26**, 393–398 (2010)
50. G.C. Matrone, C. Cipriani, M.C. Carrozza, G. Magenes, Real-time myoelectric control of a multi-fingered hand prosthesis using principal components analysis. J. Neuroeng. Rehabil. **9**, 40 (2012). https://doi.org/10.1186/1743-0003-9-40
51. M.K. Burns, K. Van Orden, V. Patel, R. Vinjamuri, Towards a wearable hand exoskeleton with embedded synergies. In *Proc Annu Int Conf IEEE Eng Med Biol Soc EMBS* (2017b), pp. 213–216. https://doi.org/10.1109/EMBC.2017.8036800
52. M. Xiloyannis, L. Cappello, D.B. Khanh, et al., Modelling and design of a synergy-based actuator for a tendon-driven soft robotic glove. Proc. IEEE RAS EMBS Int. Conf. Biomed. Robot. Biomechatronics **2016**, 1213–1219 (2016). https://doi.org/10.1109/BIOROB.2016.7523796
53. P.M. Aubin, H. Sallum, C. Walsh, et al., A pediatric robotic thumb exoskeleton for at-home rehabilitation: The Isolated Orthosis for Thumb Actuation (IOTA). IEEE Int. Conf. Rehabil. Robot. **2013**, 6650500 (2013). https://doi.org/10.1109/ICORR.2013.6650500
54. C.E. Clauser, J.T. McConville, J.W. Young, Weight, Volume, and Center of Mass of Segments of the Human Body (1969)
55. E. Lundström, A. Terént, J. Borg, Prevalence of disabling spasticity 1 year after first-ever stroke. Eur. J. Neurol. **15**, 533–539 (2008). https://doi.org/10.1111/j.1468-1331.2008.02114.x
56. P.P. Urban, T. Wolf, M. Uebele, et al., Occurrence and clinical predictors of spasticity after ischemic stroke. Stroke **41**, 2016–2020 (2010). https://doi.org/10.1161/STROKEAHA.110.581991
57. J. Wissel, L.D. Schelosky, J. Scott, et al., Early development of spasticity following stroke: a prospective, observational trial. J. Neurol. **257**, 1067–1072 (2010). https://doi.org/10.1007/s00415-010-5463-1
58. J.R. McGuire, R.L. Harvey, The prevention and management of complications after stroke. Phys. Med. Rehabil. Clin. N. Am. **10**, 857 (1999)
59. C.Y. Brown, H.H. Asada, Inter-finger coordination and postural synergies in robot hands via mechanical implementation of principal components analysis. In *IEEE International Conference on Intelligent Robots and Systems* (2007), pp. 2877–2882
60. S. Li, X. Sheng, H. Liu, X. Zhu, Design of a myoelectric prosthetic hand implementing postural synergy mechanically. Ind. Robot. Int. J. **41**, 447–455 (2014). https://doi.org/10.1108/IR-03-2014-0312
61. M. Xiloyannis, L. Galli, D. Chiaradia, et al., A soft tendon-driven robotic glove: preliminary evaluation. Biosyst. Biorobotics **21**, 329–333 (2019). https://doi.org/10.1007/978-3-030-01845-0_66
62. W. Chen, C. Xiong, S. Yue, Mechanical implementation of kinematic synergy for continual grasping generation of anthropomorphic hand. IEEE/ASME Trans. Mechatronics **20**, 1249–1263 (2015). https://doi.org/10.1109/TMECH.2014.2329006
63. M.G. Catalano, G. Grioli, E. Farnioli, et al., Adaptive synergies for the design and control of the Pisa/IIT SoftHand. Int. J. Robot. Res. **33**, 768–782 (2014). https://doi.org/10.1177/0278364913518998
64. C.D. Santina, G. Grioli, M. Catalano, et al., Dexterity augmentation on a synergistic hand: the Pisa/IIT SoftHand+. In *IEEE-RAS International Conference on Humanoid Robots* (2015), pp. 497–503
65. K. Xu, H. Liu, Y. Du, X. Zhu, Design of an underactuated anthropomorphic hand with mechanically implemented postural synergies. Adv. Robot. **28**, 1459–1474 (2014). https://doi.org/10.1080/01691864.2014.958534

66. K. Xu, Z. Liu, B. Zhao, et al., Composed continuum mechanism for compliant mechanical postural synergy: an anthropomorphic hand design example. Mech. Mach. Theory **132**, 108–122 (2019). https://doi.org/10.1016/j.mechmachtheory.2018.08.015
67. M.K. Burns, D. Pei, R. Vinjamuri, Dynamic control of virtual hand grasp using spatiotemporal synergies. *IEEE Access* (2019)

Model Predictive Control-Based Knee Actuator Allocation During a Standing-Up Motion with a Powered Exoskeleton and Functional Electrical Stimulation

Xuefeng Bao, Vahidreza Molazadeh, Albert Dodson, and Nitin Sharma

Abstract In this paper a lower-limb powered exoskeleton is combined with functional electrical stimulation of the quadriceps muscle to achieve a standing-up motion. As two actuation mechanisms (FES and the motors) act on the knee joints, it is desirable to optimally coordinate them. A feedback controller that stabilizes the desired standing-up motion is derived. The knee torques, computed by the feedback controller, are further distributed to FES and the knee electric motors by using a ratio allocation that is solved via a model predictive control method. The optimization method relies on a fatigue dynamical model. Simulations and the experimental results of the ratio allocation approach are reported for the standing-up motion.

Keywords Powered exoskeleton · Functional electrical stimulation · Nonlinear model predictive control · Sitting-to-standing motion

---

Xuefeng Bao, Vahidreza Molazadeh, Albert Dodson, and Nitin Sharma are members of IEEE.

Research reported in this chapter is supported in part by Eunice Kennedy Shriver National Institute of Child Health and Human Development of the National Institutes of Health under award number: R03HD086529. The content is solely the responsibility of the author(s) and does not necessarily represent the official views of the National Institutes of Health.

X. Bao
Department of Biomedical Engineering, Case Western Reserve University, Cleveland, OH, USA

V. Molazadeh
Department of Mechanical Engineering and Materials Science, University of Pittsburgh, Pittsburgh, PA, USA
e-mail: vam50@pitt.edu

A. Dodson · N. Sharma (✉)
Joint Department of Biomedical Engineering, North Carolina State University and University of North Carolina Chapel-Hill, Chapel Hill, NC, USA
e-mail: aedodson@ncsu.edu; nsharm23@ncsu.edu; NIS62@pitt.edu

R. Vinjamuri (ed.), *Advances in Motor Neuroprostheses*,
https://doi.org/10.1007/978-3-030-38740-2_6

1 Introduction

Achieving standing-up motion for persons with spinal cord injury (SCI) assumes significance as it is a precursor to walking. In recent years, functional electrical stimulation (FES) has been primarily used to reanimate paralyzed muscles during this motion [1–9]. FES applies external electrical currents to artificially stimulate the paralyzed muscles that generate desired limb torques [1, 10, 11]. However, a rapid onset of muscle fatigue during FES impedes an effective FES control of sitting-to-standing movements [12].

Alternatively, lower-limb powered exoskeletons can also be used to achieve sitting-to-standing motions [2, 4, 9, 13, 14]. Exoskeleton has already been shown to achieve walking [15–21], and unlike FES they can reliably generate torques that assist lower-limb movements to work for longer duration. However, powered exoskeletons may require large power source and actuators to operate, which are likely to reduce their wearability.

Hybrid devices that combine FES and powered exoskeleton have the potential to overcome rapid muscle fatigue caused by FES and also reduce the size and weight of the powered exoskeletons. Controllers that coordinate FES and the electric motors in the hybrid device have been studied in recent literature [14, 22–27]. In [28], FES and active actuators were allocated to achieve a knee extension motion with an adaptive gain-based controller, where motors were controlled by a PD controller while the FES input was provided according to an adaptive gain proportional to the motor current input. In [23], a cooperative knee joint controller was proposed for controlling a hybrid knee-ankle-foot exoskeleton, where the stimulation amount is adapted based on a fatigue estimator that measures the torque-time integral. In [24], another cooperative control approach had been developed to coordinate hip motors with FES on the hamstrings and knee motors with FES on quadriceps muscle, where the FES control was modified by the difference between the estimated muscle torque and the reference torque profile. To better allocate the actuators according to the fatigue level, a muscle fatigue dynamic model [29] was adopted in [25], where a model predictive control (MPC) method [25, 30–32] was used to achieve a leg extension motion on hybrid neuroprosthesis.

However, actuator allocation approaches in a hybrid device to achieve a standing-up motion have been minimally reported. One of the seminal papers that report patient-driven standing-up [5, 6] used only FES for actuation, where the stimulation current is determined by minimizing the upper-arm effort. In [9, 13], only a lower-limb powered exoskeleton was used to assist a user to achieve the standing-up motion. In [33] a lower-limb powered exoskeleton was combined with FES to achieve a user-driven standing-up motion. However, FES amplitude was set to be constant and was not varied dynamically. In [14], a PID controller was used to drive the exoskeleton's electric motors, while FES was triggered when a user lifts off from a seated position. Again in this work FES was not varied dynamically but was triggered based on a set knee joint angle during an extension phase. Because FES-induced fatigue is a major factor that needs to be considered, an optimal allocation of FES and the electric motors, based on the FES-induced fatigue, is desirable.

A hybrid walking neuroprosthesis device that was recently developed in our lab [33]. The exoskeleton has electrical motors mounted on the knee and hip joints while FES is used to stimulate the quadriceps muscles to assist during knee extension. In this paper a control scheme that can optimally allocate FES and the knee electric motors of the exoskeleton during the standing-up motion is proposed. In the scheme, a robust feedback controller is derived to provide stabilizing control signals. The feedback control computes the total torques required for both knee and hip joints to track a desired standing-up motion. Then, an online optimal control method is used to determine an optimal ratio between the FES-induced torque to the motor torque at the knee joint. The optimization uses a normalized fatigue level that is predicted by a dynamic muscle fatigue model to determine the actuator allocation. This allows dynamic allocation of control inputs to FES and the electric motors. Experiments were performed on an able-bodied participant to validate the proposed control method.

2 System Dynamics

A 2 degree of freedom lower-limb dynamics during the standing-up motion, as in Fig. 1, can be expressed as

$$M(\boldsymbol{q})\ddot{\boldsymbol{q}} + C_m(\boldsymbol{q},\dot{\boldsymbol{q}})\dot{\boldsymbol{q}} + F(\boldsymbol{q},\dot{\boldsymbol{q}}) + G(\boldsymbol{q}) + \boldsymbol{\omega} = T \tag{1}$$

where the terms $\boldsymbol{q}, \dot{\boldsymbol{q}}, \ddot{\boldsymbol{q}} \in \mathbb{R}^2$ are the joint angular position vector, angular velocity vector , and angular acceleration vector, respectively. The vector $\boldsymbol{q} = \begin{bmatrix} q_1 \ q_2 \end{bmatrix}^T$ contains the knee joint angular position, q_1 and the hip joint angular position, q_2. The torque vector

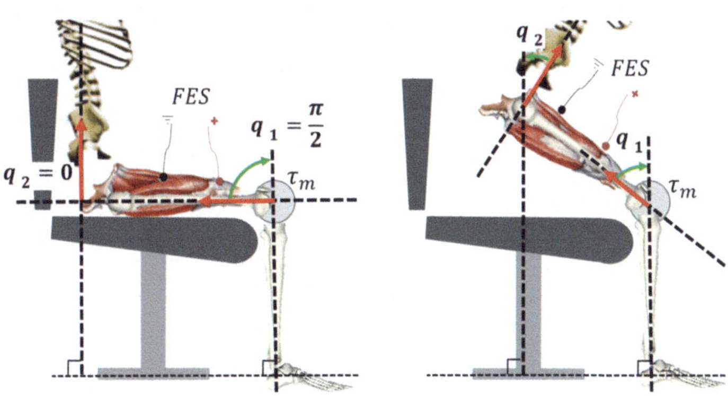

Fig. 1 A representative figure depicting the hybrid neuroprosthesis for standing-up motion

$$\boldsymbol{T} \equiv \left[T_1\ T_2\right]^T = \left[\tau_k\ \tau_h\right]^T + \left[\tau_a\ 0\right]^T \tag{2}$$

denotes the total torque vector that contains the knee motor torque $\tau_k \in \mathbb{R}$, hip motor torque $\tau_h \in \mathbb{R}$, and $\tau_a(q_1, \dot{q}_1, \mu, u_{fes}) \in \mathbb{R}$ denotes the FES-induced torque at the knee joint. $u_{fes} \in \mathbb{R}$ the normalized FES current level, and $\mu \in \mathbb{R}$ the muscle fatigue [5, 29]. In (1) we can see that only knee joint is stimulated by FES. The disturbance term $\boldsymbol{\omega} \in \mathbb{R}^2$ is bounded as $|\boldsymbol{\omega}| \leq \boldsymbol{\Omega} \in \mathbb{R}^2$. The matrix $M(\boldsymbol{q}) \in \mathbb{R}^{2\times2}$ denotes the moment of inertia, which is symmetric, positive definite, bounded, and invertible. The matrix $C_m(\boldsymbol{q}, \dot{\boldsymbol{q}}) \in \mathbb{R}^{2\times2}$ denotes the Coriolis term [34]. The vector $G(\boldsymbol{q}) \in \mathbb{R}^2$ denotes the gravity vector and $F(\boldsymbol{q}, \dot{\boldsymbol{q}}) \in \mathbb{R}^2$ denotes the passive muscle torque vector. The equation of motion (1) can be rewritten as

$$\ddot{\boldsymbol{q}} + \Psi\left(\boldsymbol{q}, \dot{\boldsymbol{q}}\right) + \boldsymbol{v} = B\left(\boldsymbol{q}\right)\boldsymbol{T} \tag{3}$$

where $\Psi\left(\boldsymbol{q}, \dot{\boldsymbol{q}}\right) = M^{-1}C + M^{-1}F + M^{-1}G$, $B\left(\boldsymbol{q}\right) = M^{-1}$, $\boldsymbol{v} \equiv M^{-1}\boldsymbol{\omega}$.

3 Standing Motion Planning

The joint angle, velocity, and acceleration can be represented in an output vector. Let us define $\boldsymbol{h}_d = \left[q_{d_1}, \dot{q}_{d_1}, \ddot{q}_{d_1}, q_{d_2}, \dot{q}_{d_2}, \ddot{q}_{d_2}\right]^T$ as the reference output vector (represents the optimal motion) and $\boldsymbol{h} = \left[q_1, \dot{q}_1, \ddot{q}_1, q_2, \dot{q}_2, \ddot{q}_2\right]^T$ as the actual output vector. The reference output can be computed using a virtual constraint approach [35], which is obtained by substituting the independent join angle function, $\theta\left(\boldsymbol{q}, \dot{\boldsymbol{q}}\right) \in \mathbb{R}$

$$\theta\left(\boldsymbol{q}, \dot{\boldsymbol{q}}\right) = \zeta_1 q_1 + \zeta_2 \dot{q}_1 + \zeta_3 q_2 + \zeta_4 \dot{q}_2$$

where $\forall i = 1, 2, 3, 4$ $\zeta_{\mathrm{i}} \in \mathbb{R}$ are chosen such that $\theta\left(\boldsymbol{q}, \dot{\boldsymbol{q}}\right)$ is monotonically increasing. Then Bezier polynomials can be used to obtain the reference output

$$\boldsymbol{q}_d o \theta\left(\boldsymbol{q}, \dot{\boldsymbol{q}}\right) = \begin{bmatrix} b_1 o \varsigma(\boldsymbol{q}, \dot{\boldsymbol{q}}) \\ b_2 o \varsigma(\boldsymbol{q}, \dot{\boldsymbol{q}}) \end{bmatrix} \tag{4}$$

where

$$b_i(w) = \sum_{k=0}^{M} \varrho_k^i \frac{N_b!}{k!(N_b - k)!} w^k (1 - w)^{M-k}. \tag{5}$$

In (5), M, $N_b \in \mathbb{I}^+$ denote the order and number of Bezier polynomial terms, respectively, ϱ_k^i is an optimization parameter, and w is obtained according to

$$w(\boldsymbol{q}) = \frac{\theta(\boldsymbol{q}, \dot{\boldsymbol{q}}) - \theta^+}{\theta^- - \theta^+} \tag{6}$$

where $\theta^-$and $\theta^+$are the lower and upper limitations of the $\theta(\boldsymbol{q}, \dot{\boldsymbol{q}})$, respectively [36].

It can be seen that in the design, the reference output is time invariant; i.e., it is constrained only to the actual joint kinematics and there exists a motion generation law $\Theta(\cdot, \cdot)$ such that $\boldsymbol{h}_d = \Theta(\boldsymbol{q}, \dot{\boldsymbol{q}})$.

4 Feedback Control Development

The control objective is to minimize the error term $\boldsymbol{e} = \boldsymbol{q}_d - \boldsymbol{q}$, where $\boldsymbol{q}_d = \begin{bmatrix} q_{d_1} \; q_{d_2} \end{bmatrix}^T$. Define an auxiliary signal, $\boldsymbol{s} \in \mathbb{R}^2$,

$$\boldsymbol{s} = \dot{\boldsymbol{e}} + \lambda \boldsymbol{e}. \tag{7}$$

After taking time derivative of (7) and on substituting (3),

$$\dot{\boldsymbol{s}} = \ddot{\boldsymbol{q}}_d + \lambda \dot{\boldsymbol{q}}_d - \lambda \dot{\boldsymbol{q}} + \Psi + \boldsymbol{v} - B\boldsymbol{T}. \tag{8}$$

Based on the subsequent stability analysis, a feedback law $K(\cdot, \cdot)$ is designed such that

$$\begin{aligned} \boldsymbol{T} = \hat{M}\left(\ddot{\boldsymbol{q}}_d + \lambda \dot{\boldsymbol{q}}_d - \lambda \boldsymbol{q} + \kappa_2 \boldsymbol{s}\right) + \hat{C}_m \boldsymbol{s} + \hat{C} + \hat{F} + \hat{G} \\ + \kappa_1 \mathrm{sgn}(\boldsymbol{s}) + \kappa_0\left(\boldsymbol{e}, \dot{\boldsymbol{e}}\right)\boldsymbol{s} + \Gamma \mathrm{sgn}\left(\dot{\boldsymbol{q}}^T \boldsymbol{s}\right)\dot{\boldsymbol{q}} + \hat{\boldsymbol{\omega}} \end{aligned} \tag{9}$$

where κ_0 is a positive and monotonically increasing function, $\kappa_1 \in \mathbb{R}^+$ and $\kappa_2 \in \mathbb{R}^+$ are control gains, $\Gamma \in \mathbb{R}^{2\times 2}$ is a positive definite control gain matrix, and $\hat{M} \in \mathbb{R}^{2\times 2}$, $\hat{C}_m \in \mathbb{R}^{2\times 2}$, $\hat{C} \in \mathbb{R}^2$, $\hat{F} \in \mathbb{R}^2$, $\hat{G} \in \mathbb{R}^2$, $\hat{\omega} \in \mathbb{R}^2$ as estimates of M, C_m, C, F, G, ω , respectively, where $C(\boldsymbol{q}, \dot{\boldsymbol{q}}) = C_m(\boldsymbol{q}, \dot{\boldsymbol{q}})\dot{\boldsymbol{q}}$ [34]. The controller is expressed in the following compact form:

$$\boldsymbol{T} = K(\Theta(\boldsymbol{q}, \dot{\boldsymbol{q}}), \boldsymbol{h}). \tag{10}$$

On substituting (9) into (8)

$$\begin{aligned} \dot{\boldsymbol{s}} =& \Psi - M^{-1}C - M^{-1}F - M^{-1}G - M^{-1}C_m\boldsymbol{s} \\ &+ M^{-1}\tilde{M}\left(\ddot{\boldsymbol{q}}_d + \lambda\dot{\boldsymbol{q}}_d\right) + M^{-1}\tilde{C} + M^{-1}\tilde{F} \\ &+ M^{-1}\tilde{G} + \boldsymbol{v} - M^{-1}\hat{\boldsymbol{\omega}} - \kappa_1 M^{-1}\mathrm{sgn}(\boldsymbol{s}) \\ &+ \kappa_2 M^{-1}\tilde{M}\boldsymbol{s} + M^{-1}\tilde{C}_m\boldsymbol{s} - \kappa_0 M^{-1}\boldsymbol{s} \\ &- M^{-1}\Gamma\mathrm{sgn}(\dot{\boldsymbol{q}}^T\boldsymbol{s})\dot{\boldsymbol{q}} - \lambda M^{-1}\tilde{M}\dot{\boldsymbol{q}} - \kappa_2\boldsymbol{s} \end{aligned}$$

where $\tilde{M} = M - \hat{M} = \begin{bmatrix} \tilde{m}_{11} & \tilde{m}_{12} \\ \tilde{m}_{21} & \tilde{m}_{22} \end{bmatrix}$, $\tilde{C}_m = C_m - \hat{C}_m$, $\tilde{C} = C - \hat{C}$, $\tilde{F} = F - \hat{F}$, and $\tilde{G} = G - \hat{G}$, $\tilde{\boldsymbol{\omega}} = \boldsymbol{\omega} - \hat{\boldsymbol{\omega}}$, $\tilde{\boldsymbol{v}} = \boldsymbol{v} - \hat{\boldsymbol{v}}$, $\hat{\boldsymbol{v}} = M^{-1}\hat{\boldsymbol{\omega}}$.

It is further simplified to

$$\begin{aligned} M\dot{\boldsymbol{s}} = & - C_m\boldsymbol{s} + \tilde{\boldsymbol{n}} + \tilde{\boldsymbol{y}} - \kappa_1 \mathrm{sgn}(\boldsymbol{s}) - \lambda\tilde{M}\dot{\boldsymbol{q}} \\ & - \Gamma \mathrm{sgn}\left(\dot{\boldsymbol{q}}^T\boldsymbol{s}\right)\dot{\boldsymbol{q}} + \Upsilon\boldsymbol{s} - \kappa_0\boldsymbol{s} - \kappa_2 M\boldsymbol{s}. \end{aligned} \tag{11}$$

In (11) $\Upsilon \in \mathbb{R}^{2\times 2}$ is defined as $\Upsilon \triangleq \begin{bmatrix} \varepsilon_1 & \varepsilon_2 \\ \varepsilon_3 & \varepsilon_4 \end{bmatrix} = \kappa_2\tilde{M} + \tilde{C}_m$, where $\varepsilon_i \in \mathbb{R}\forall i = 1, 2, 3, 4$, $\tilde{\boldsymbol{y}} \in \mathbb{R}^2$ is defined as $\tilde{\boldsymbol{y}} = \tilde{C} + \tilde{F} + \tilde{G}$, $\tilde{\boldsymbol{n}} \in \mathbb{R}^2$ is defined as $\tilde{\boldsymbol{n}} \triangleq [\tilde{n}_1, \tilde{n}_2]^T = \tilde{\boldsymbol{\omega}} + \tilde{\boldsymbol{d}}$, and $\tilde{\boldsymbol{d}} \in \mathbb{R}^2$ is defined as $\tilde{\boldsymbol{d}} \equiv \tilde{M}\ddot{\boldsymbol{q}}_d + \lambda\tilde{M}\dot{\boldsymbol{q}}_d$. It is also assumed that $\tilde{\boldsymbol{y}}$ is bounded as[34]

$$\|\tilde{\boldsymbol{y}}\| \leq \tilde{Y}(\|\boldsymbol{e}, \dot{\boldsymbol{e}}\|^2)$$

where $\tilde{Y}$ is a positive and monotonically increasing function.

Theorem 1 *The control law in (9) makes the error dynamical system in (11) semi-globally exponentially stable, provided the following gain conditions hold true:*

$$\begin{aligned} & \kappa_0 > \max\{|\varepsilon_i|; \forall i = 1, 2, 3, 4\} + \tilde{Y}(\|\boldsymbol{e}, \dot{\boldsymbol{e}}\|^2) \\ & \kappa_1 > \max\{|\tilde{n}_1|, |\tilde{n}_2|\} \\ & \Gamma = \begin{bmatrix} \gamma_{11} & \gamma_{12} \\ \gamma_{21} & \gamma_{22} \end{bmatrix} \forall i, j\, \gamma_{ij} \geq \lambda\tilde{m}_{ij}. \end{aligned} \tag{12}$$

Proof Define a positive definite Lyapunov candidate, $V \in \mathbb{R}$

$$V = \frac{1}{2}\boldsymbol{s}^T M\boldsymbol{s} \tag{13}$$

such that $\lambda_m \|\boldsymbol{s}\|^2 \leq V \leq \lambda_M \|\boldsymbol{s}\|^2$, where λ_m and λ_M are the minimum and maximum eigenvalue of M, respectively.

The time derivative of V is

$$\dot{V} = \boldsymbol{s}^T M\dot{\boldsymbol{s}} + \frac{1}{2}\boldsymbol{s}^T \dot{M}\boldsymbol{s}. \tag{14}$$

On substituting (11) into (14), and by applying the skew symmetric property [34], i.e., $\dot{M} - 2C_m = 0$, we can obtain

$$\begin{aligned} \dot{V} = & - \kappa_1\boldsymbol{s}^T \mathrm{sgn}(\boldsymbol{s}) + \boldsymbol{s}^T\tilde{\boldsymbol{n}} + \boldsymbol{s}^T\tilde{\boldsymbol{y}} + \boldsymbol{s}^T\Upsilon\boldsymbol{s} - \boldsymbol{s}^T\kappa_0\boldsymbol{s} \\ & - \lambda\boldsymbol{s}^T\tilde{M}\dot{\boldsymbol{q}} - \boldsymbol{s}^T\Gamma \mathrm{sgn}\left(\dot{\boldsymbol{q}}^T\boldsymbol{s}\right)\dot{\boldsymbol{q}} - \boldsymbol{s}^T\kappa_2 M\boldsymbol{s} \end{aligned} \tag{15}$$

$$\leq -\kappa_1 s^T \text{sgn}(s) + |s_1| |\tilde{n}_1| + |s_2| |\tilde{n}_2|$$
$$+ \|s\| \tilde{Y}(\|e, \dot{e}\|^2) + s^T \Upsilon s - s^T \kappa_0 s$$
$$- \lambda s^T \tilde{M} \dot{q} - s^T \Gamma \text{sgn}\left(\dot{q}^T s\right) \dot{q} - \kappa_2 s^T M s.$$

By applying (12), (15) can be further simplified to

$$\dot{V} < -\kappa_2 s^T M s \leq -2\kappa_2 V.$$

This means that V is semi-globally exponentially stable, which further implies that e goes to $\mathbf{0}$ exponentially. ■

5 Model Predictive Control-Based Ratio Allocation Method

The control allocation problem is to distribute T_1 in (2) among the knee torque, τ_a generated by FES and the knee electrical motor, τ_k. Further, we are interested in computing the stimulation amplitude for the quadriceps muscles that generate τ_a. Therefore, the following model is used to determine stimulation current.

5.1 *Muscle Force Generation and Fatigue Model*

The active knee torque, $\tau_a = \Phi_a\left(\phi\left(q_1, \dot{q}_1\right), \mu, u_{fes}\right) \in \mathbb{R}^+ \cup \{0\}$ is [37]

$$\tau_a = \phi\left(q_1, \dot{q}_1\right) \mu u_{fes}, \tag{16}$$

where $\phi\left(q_1, \dot{q}_1\right) = \left(c_2 q_1^2 + c_1 q_1 + c_0\right)\left(1 - c_3 \dot{q}_1\right)$ ($c_i \forall i = 0, 1, 2, 3$ are model parameters) is the torque-knee angle and knee angular velocity relationships [37], u_{fes} is the normalized stimulation amplitude, and μ is the normalized fatigue variable driven by the fatigue dynamics $\dot{\mu} = \Phi_\mu\left(\mu, u_{fes}\right) \in \mathbb{R}$

$$\dot{\mu} = \frac{\left(\mu_{min} - \mu\right) u_{fes}}{T_f} + \frac{\left(1 - \mu\right)\left(1 - u_{fes}\right)}{T_r} \tag{17}$$

5.2 *Optimization Problem*

Based on the knowledge of $\bar{T}$ ($\bar{\cdot}$ represents the nominal signals in the MPC), which is obtained by running the feedback controller (10) through the nominal model (without disturbance), an MPC approach is designed to determine the allocation

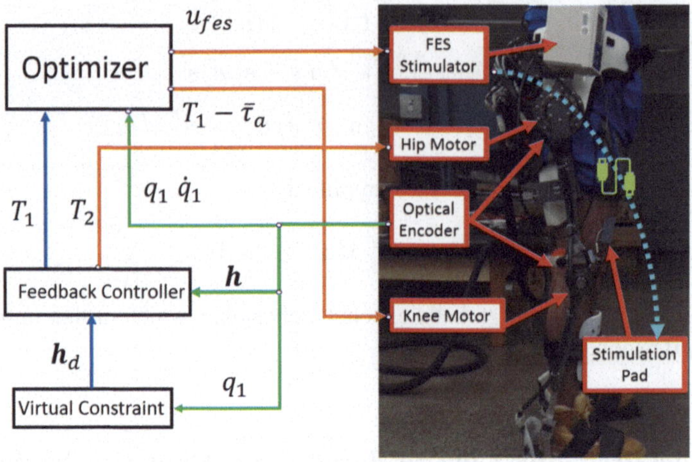

Fig. 2 The control loop of the ratio allocation approach

between the motor and FES input. The optimization is also constrained to the estimated muscle fatigue model (17). The control loop is shown in Fig. 2.

The optimization problem is

$$\min_{\bar{u}_{fes}} J(t_k) = \int_{t_k}^{t_k+T_N} \left\{ \bar{\tau}_k^2 + \frac{w_1}{\bar{\mu}+\epsilon} \bar{\tau}_a^2 + w_2(\bar{\mu} - \bar{\mu}_o)^2 \right\} dt \tag{18}$$

$$\text{s.t. } \bar{\tau}_k + \bar{\tau}_a = \bar{T}_1$$

$$\bar{\boldsymbol{T}} = K(\Theta(\bar{\boldsymbol{q}}, \dot{\bar{\boldsymbol{q}}}), \bar{\boldsymbol{h}})$$

$$M(\bar{\boldsymbol{q}})\ddot{\bar{\boldsymbol{q}}}(t) + C_m(\bar{\boldsymbol{q}}, \dot{\bar{\boldsymbol{q}}})\dot{\bar{\boldsymbol{q}}}(t) + F(\bar{\boldsymbol{q}}, \dot{\bar{\boldsymbol{q}}}) + G(\bar{\boldsymbol{q}}) = \bar{\boldsymbol{T}}$$

$$\bar{\tau}_a = \Phi_a(\phi(q_1, \dot{q}_1), \bar{\mu}, \bar{u}_{fes})$$

$$\dot{\bar{\mu}} = \Phi_\mu(\bar{\mu}, \bar{u}_{fes})$$

$$\bar{u}_{fes} \in \mathcal{U}_{fes}$$

the objective index $J(t) \in \mathbb{R}^+ \cup \{0\}$ in (18) is the cost function, $\bar{\mu}_o$ is the estimated fatigue of contralateral leg, $\mathcal{U}_{fes}$ is the input constraint (normalized as $[0, 1]$[25, 38]), $\epsilon > 0$ is a constant, and $w_{1,2} > 0$ is a user-defined weight. When the optimal solution, $u^*_{fes}(t| : t \in [t_k, t_k + T_N]) = \text{argmin}\{J(t)\}$, is found, $u_{fes} = u^*_{fes}(t| : t = t_k \to t_k + \varepsilon)$ is applied to the system, where ε is an infinitesimal time constant that makes $t_{k+1} = t_k + \varepsilon$ [31].

For the ratio allocation optimization, a gradient search algorithm was adopted [31] to solve the optimization problem. The detailed algorithm can be found in Table 1.

Table 1 Detailed MPC algorithm

1	**Initialization:** $j = 0$
(1a)	Set the convergence tolerance ε_j.
(1b)	Measure $\boldsymbol{q}(t_k), \dot{\boldsymbol{q}}(t_k)$.
(1c)	Use virtual constraint and feedback controller to get $\boldsymbol{h}_d(\tau)$, $\bar{\boldsymbol{h}}(\tau)$, and $T_1(\tau)$, where $\tau \in [t_k, t_k + T_N]$.
(1d)	Choose initial control trajectory $\bar{u}_{fes}(\tau) \in \mathcal{U}_{[t_k, t_k+T_N]}$, where $\tau \in [t_k, t_k + T_N]$.
(1e)	Use $\bar{u}_{fes}(\tau)$ and $\bar{\boldsymbol{h}}(\tau)$ to obtain $\bar{\tau}_a(\tau)$, therefore, $J^{(j)}(t_k)$, where $\tau \in [t_k, t_k + T_N]$.
2	**Optimal Solution Searching:**
(2a)	Integrate backward in time to solve for the costates $l^{(j)}(\tau)$ by minimizing the Hamiltonian $H = J_{mpc} + l^T \Phi_\mu$, so that $\dot{l}(\tau) = -\frac{\partial H(\bar{\mu}, l, \bar{u}_{fes})}{\partial \bar{\mu}}$.
(2b)	Compute the search direction, $a^{(j)}(\tau)$, from the Hamiltonian $a^{(j)}(\tau) = -\frac{\partial H(x, l, \bar{u}_{fes})}{\partial \bar{u}_{fes}}$.
(2c)	Compute the optimal step size, $\sigma^{(j)}$, with adaptive setting in [31].
(2d)	Update the control trajectory $\bar{u}_{fes}^{(j+1)}(\tau) = \psi(\bar{u}_{fes}^{(j)} + \sigma^{(j)} a^{(j)})$, where ψ denotes the constraints.
(2e)	Use $\bar{u}_{fes}^{(j+1)}$ to get $J^{(j+1)}(t_k)$.
(2f)	Check Quit Conditions (i) quit if $\left\| J^{(j+1)}(t_k) - J^{(j)}(t_k) \right\| \leq \varepsilon_j$, (ii) quit if j has exceeded the max iteration limit, N_t, (iii) otherwise set $j = j + 1$ and reiterate gradient step from (2a).

6 Results

6.1 Simulation

To validate the control method, simulations were run. The actual system parameters in the simulation are set to be slightly different than the nominal one. In the simulation, only one leg is considered by assuming that the contralateral leg dynamics is the symmetric and constrained to the primary leg. We also assume that fatigue levels of the two legs are equal, i.e., $\bar{\mu} = \bar{\mu}_o$. In the simulation, allocation results were investigated by setting $w_1 = 1$. The simulation result can be seen in Fig. 3.

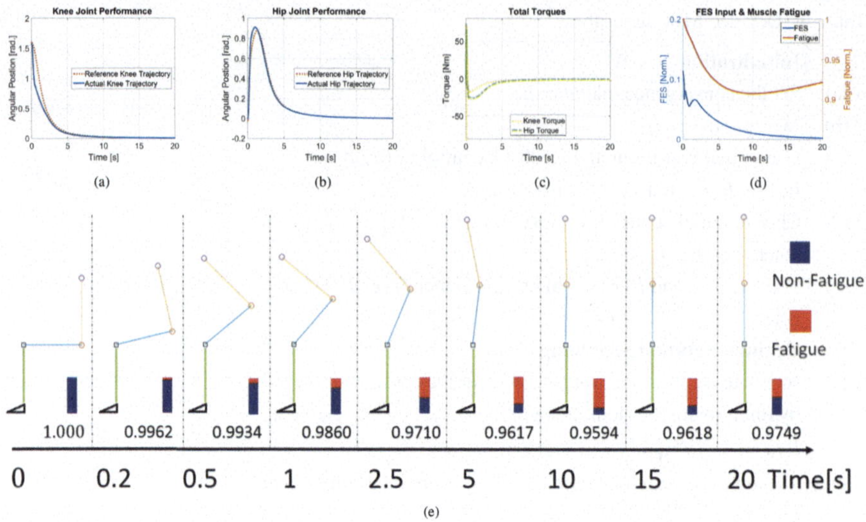

Fig. 3 This figure demonstrate the simulation results. (**a**) Knee performance. (**b**) Hip performance. (**c**) Total motor torques. (**d**) FES and fatigue. (**e**) This figure visualizes the standing-up motion and the fatigue levels corresponding to (**d**)

7 Conclusion

This paper proposed a control scheme that includes an MPC method and a feedback controller to control a hybrid neuroprosthesis that can potentially help a person with SCI to achieve a standing-up motion. In the control scheme a ratio allocation method is proposed to allocate between FES and the powered exoskeleton. Simulation and experimental results validated the method. In our future work, the proposed method will be investigated on a person with SCI.

References

1. H. Chizek, R. Kobetic, E. Marsolais, J. Abbas, I. Donner, E. Simon, Control of functional neuromuscular stimulation systems for standing and locomotion in paraplegics. Proc. IEEE **76**(9), 1155–1165 (1988)
2. N. Donaldson, C.-N. Yu, Fes standing: control by handle reactions of leg muscle stimulation (CHRELMS). IEEE Trans. Rehabil. Eng. **4**(4), 280–284 (1996)
3. N. Donaldson, C.-H. Yu, A strategy used by paraplegics to stand up using fes. IEEE Trans. Rehabil. Eng. **6**(2), 162–167 (1998)
4. H. Kagaya, M. Sharma, R. Kobetic, E.B. Marsolais, Ankle, knee, and hip moments during standing with and without joint contractures: simulation study for functional electrical stimulation. Am. J. Phys. Med. Rehabil. **77**(1), 49–54 (1998)
5. R. Riener, T. Fuhr, Patient-driven control of FES-supported standing up: a simulation study. IEEE Trans. Rehabil. Eng. **6**, 113–124 (1998)

6. R. Riener, M. Ferrarin, E.E. Pavan, C.A. Frigo, Patient-driven control of fes-supported standing up and sitting down: experimental results. IEEE Trans. Rehabil. Eng. **8**(4), 523–529 (2000)
7. R. Kobetic, C. To, J. Schnellenberger, M. Audu, T. Bulea, R. Gaudio, G. Pinault, S. Tashman, R. Triolo, Development of hybrid orthosis for standing, walking, and stair climbing after spinal cord injury. J. Rehabil. Res. Dev. **46**(3), 447–462 (2009)
8. S. Jatsun, S. Savin, A. Yatsun, R. Turlapov, Adaptive control system for exoskeleton performing sit-to-stand motion, in *2015 10th International Symposium on Mechatronics and Its Applications (ISMA)* (IEEE, 2015), pp. 1–6
9. W. Huo, S. Mohammed, Y. Amirat, K. Kong, Active impedance control of a lower limb exoskeleton to assist sit-to-stand movement, in *IEEE International Conference on Robotics and Automation (ICRA)* (IEEE, 2016), pp. 3530–3536
10. R. Kobetic, B. Marsolais, Synthesis of paraplegic gait with multichannel functional neuromuscular stimulation. IEEE Trans. Rehabil. Eng. **2**(2), 66–79 (1994)
11. W. Durfee, Gait restoration by functional electrical stimulation, in *Climbing and Walking Robots* (Springer, Berlin, 2006), pp. 19–26
12. N. Sharma, N.A. Kirsch, N.A. Alibeji, W.E. Dixon, A non-linear control method to compensate for muscle fatigue during neuromuscular electrical stimulation. Front. Robot. AI **4**, 68 (2017)
13. S. Mefoued, S. Mohammed, Y. Amirat, G. Fried, Sit-to-stand movement assistance using an actuated knee joint orthosis, in *2012 4th IEEE RAS & EMBS International Conference on Biomedical Robotics and Biomechatronics (BioRob)* (IEEE, 2012), pp. 1753–1758
14. M.A. Alouane, W. Huo, H. Rifai, Y. Amirat, S. Mohammed, Hybrid fes-exoskeleton controller to assist sit-to-stand movement. *IFAC-PapersOnLine* **51**(34), 296–301 (2019)
15. H. Kawamoto, S. Kanbe, Y. Sankai, Power assist method for hal-3 estimating operator's intention based on motion information, in *The 12th IEEE International Workshop on Robot and Human Interactive Communication, 2003. Proceedings. ROMAN 2003* (IEEE, 2003), pp. 67–72
16. R. Farris, H. Quintero, M. Goldfarb, Preliminary evaluation of a powered lower limb orthosis to aid walking in paraplegic individuals. IEEE Trans. Neural Syst. Rehabil. Eng. **19**(6), 652–659 (2011)
17. P. Neuhaus, J. Noorden, T. Craig, T. Torres, J. Kirschbaum, J. Pratt, Design and evaluation of mina: a robotic orthosis for paraplegics, in *IEEE ICORR* (2011), pp. 1–8
18. K. Strausser, H. Kazerooni, The development and testing of a human machine interface for a mobile medical exoskeleton, in *IEEE/RSJ IROS* (2011), pp. 4911–4916
19. A. Esquenazi, M. Talaty, A. Packel, M. Saulino, The ReWalk powered exoskeleton to restore ambulatory function to individuals with thoracic-level motor-complete spinal cord injury. Am. J. Phys. Med. Rehabil. **91**(11), 911–921 (2012)
20. E. Strickland, Good-bye, wheelchair. IEEE Spectr. **49**(1), 30–32 (2012)
21. A.J. Young, D.P. Ferris, State of the art and future directions for lower limb robotic exoskeletons. IEEE Trans. Neural Syst. Rehabil. Eng. **25**(2), 171–182 (2017)
22. H. Quintero, R. Farris, W. Durfee, M. Goldfarb, Feasibility of a hybrid-FES system for gait restoration in paraplegics, in *IEEE EMBC* (2010), pp. 483–486
23. A. del Ama, Á. Gil-Agudo, J. Pons, J. Moreno, Hybrid FES-robot cooperative control of ambulatory gait rehabilitation exoskeleton. J. NeuroEng. Rehabil. **11**(1), 27 (2014)
24. K. Ha, S. Murray, M. Goldfarb, An approach for the cooperative control of FES with a powered exoskeleton during level walking for persons with paraplegia. IEEE Trans. Neural Syst. Rehabil. Eng. **24**, 455–466 (2016)
25. N. Kirsch, X. Bao, N. Alibeji, B. Dicianno, N. Sharma, Model-based dynamic control allocation in a hybrid neuroprosthesis. IEEE Trans. Neural Syst. Rehabil. Eng. **26**(1), 224–232 (2018)
26. N.A. Alibeji, V. Molazadeh, B.E. Dicianno, N. Sharma, A control scheme that uses dynamic postural synergies to coordinate a hybrid walking neuroprosthesis: theory and experiments. Front. Neurosci. **12**, 159 (2018)

27. N.A. Alibeji, V. Molazadeh, F. Moore-Clingenpeel, N. Sharma, A muscle synergy-inspired control design to coordinate functional electrical stimulation and a powered exoskeleton: artificial generation of synergies to reduce input dimensionality. IEEE Control Syst. Mag. **38**(6), 35–60 (2018)
28. H. Quintero, R. Farris, K. Ha, M. Goldfarb, Preliminary assessment of the efficacy of supplementing knee extension capability in a lower limb exoskeleton with FES. IEEE Eng. Med. Biol. Soc. **2012**, 3360–3363 (2012)
29. R. Riener, J. Quintern, G. Schmidt, Biomechanical model of the human knee evaluated by neuromuscular stimulation. J. Biomech. **29**, 1157–1167 (1996)
30. D.Q. Mayne, J.B. Rawlings, C.V. Rao, P.O. Scokaert, Constrained model predictive control: stability and optimality. Automatica **36**(6), 789–814 (2000)
31. K. Graichen, B. Käpernick, A real-time gradient method for nonlinear model predictive control. INTECH Open Access Publisher (2012)
32. J.B. Rawling, D.Q. Mayne, M.M. Diehl, Model Predictive Control: Theory, Computation, and Design, 2nd edn. (Nob Hill Publishing, LLC, Madison, 2017)
33. A. Dodson, A novel user-controlled assisted standing control system for a hybrid neuroprosthesis. Master's thesis, University of Pittsburgh, 2018
34. F. Lewis, D. Dawson, C. Abdallah, *Robot Manipulator Control: Theory and Practice* (CRC Press, Boca Raton, 2003)
35. E.R. Westervelt, J.W. Grizzle, C. Chevallereau, J.H. Choi, B. Morris, *Feedback Control of Dynamic Bipedal Robot Locomotion*, vol. 28 (CRC Press, Boca Raton, 2007)
36. V. Molazadeh, Z. Sheng, X. Bao, N. Sharma, A robust iterative learning switching controller for following virtual constraints: application to a hybrid neuroprosthesis. IFAC-PapersOnLine **51**(34), 28–33 (2019)
37. N. Kirsch, N. Alibeji, B.E. Dicianno, N. Sharma, Switching control of functional electrical stimulation and motor assist for muscle fatigue compensation, in *American Control Conference (ACC), 2016* (IEEE, 2016), pp. 4865–4870
38. Z. Sun, X. Bao, N. Sharma, Lyapunov-based model predictive control of an input delayed functional electrical simulation. IFAC-PapersOnLine **51**(34), 290–295 (2019)

Deep Brain Stimulation for Gait and Postural Disturbances in Parkinson's Disease

Hanyan Li and George C. McConnell

Abstract Parkinson's disease (PD) is a progressive neurodegenerative disorder characterized by distal (i.e., tremor, bradykinesia, and rigidity) and axial motor symptoms (i.e., gait and postural disturbances). Deep brain stimulation (DBS) is a neurosurgical approach that uses electrical current delivered by an implantable pulse generator to modulate neural activity. Although DBS at the subthalamic nucleus (STN) and the internal globus pallidus (GPi) are well established for the treatment of the distal symptoms in PD, long-term studies of axial symptoms show a decline in efficacy with progression of the disease. Currently, there is no pharmacological or neurosurgical treatment available for the axial symptoms of advanced PD. Thus, the design of novel stimulation strategies to treat gait disturbances and postural instability has been investigated, including targets such as the pedunculopontine nucleus (PPN) and the substantia nigra pars reticulata (SNr). Here, we reviewed the current state of understanding regarding the effects of STN/GPi DBS, PPN DBS, and SNr DBS on gait and postural disturbances in PD and the proposed underlying mechanisms of action. The stimulation parameters (i.e., location, frequency, amplitude, and pulse width) and localization criteria for accurate placement of DBS electrodes within each target are discussed. As DBS at spatially distinct subregions of a target impacts the effectiveness of the therapy, electrode misplacement may directly contribute to the mixed results of DBS on the gait and postural disturbances of PD. We highlight the need for future studies to provide details on the specific subregion of the stimulation target to further advance the field.

Keywords Parkinson's disease (PD) · Deep brain stimulation (DBS) · Subthalamic nucleus (STN) · Substantia nigra pars reticulata (SNr) · Pedunculopontine nucleus (PPN) · Gait disturbances · Postural instability

H. Li · G. C. McConnell (✉)
Department of Biomedical Engineering, Stevens Institute of Technology, Hoboken, NJ, USA
e-mail: gmcconne@stevens.edu

R. Vinjamuri (ed.), *Advances in Motor Neuroprostheses*,
https://doi.org/10.1007/978-3-030-38740-2_7

1 Introduction

1.1 Parkinson's Disease

Parkinson's disease (PD) is one of the most common neurodegenerative disorders. There were 680,000 individuals in the USA with PD in 2010, and that number will rise to approximately 930,000 in 2020 and 1,238,000 in 2030 [1]. PD has a major impact on the quality of life of patients, their families, and caregivers carrying a significant social economic burden, which has been estimated in a recent study to exceed $14.4 billion in 2010 (approximately $22,800 per patient [2]). PD is characterized by a progressive degeneration of dopaminergic neurons in the substantia nigra pars compacta (SNc), which produce dopamine, a chemical that controls movement and coordination [3]. Although the specific symptoms vary from person to person, the primary signs of PD include blank facial expression, slowing monotonous slurred speech, rigidity and tremor of extremities and head, forward tilt of posture, reduced arm swinging, and shuffling gait [4].

1.2 Gait and Postural Disturbances in PD

Parkinsonian gait disorders consist of reduced walking speed and step length, start hesitation, freezing, and festination [5]. Festination gait is described as rapid small steps done in an attempt to keep the center of gravity in between the feet while the trunk is leaning forward involuntarily. Freezing of gait (FOG) is an episodic transient disruption of gait that typically lasts a few seconds and is associated with a unique sensation: patients feels that their feet are glued to the ground, causing them to remain in place despite making a concerted effort to overcome the motor block and move forward [6]. When FOG occurs during gait initiation, it is characterized by repeated ineffective anticipatory postural adjustments and leads to a failure of gait initiation and sometimes to a fall. FOG may also occur while patients are walking. There is an abrupt decrease of step length, increase of step frequency, and step-to-step variability that precede a complete blockade of gait and falls. Postural disturbances related with hypertonia are also described in PD gait. Postural instability compromises the ability to maintain balance during everyday tasks and can result in falls, which constitute a major public health problem [7–9]. Almost 50% of falls occur during walking, in particular during the initiation and termination of gait [10], and the likelihood of falls is ~60%, which is higher than healthy elderly [11]. Falls and freezing of gait are responsible for high morbidity (e.g., fractures, residential health care) and increased significantly mortality [11]. The most widely used clinical rating scale for the severity of PD symptoms is the Unified Parkinson's disease rating scale (UPDRS) [12].

1.3 Treatment Options for Gait and Postural Disturbances in PD

Levodopa (L-DOPA), a dopaminergic drug, remains the gold-standard treatment for PD, although treatment is complicated by L-DOPA-induced dyskinesia (LID) in later stages [13]. Deep brain stimulation (DBS) offers a powerful therapeutic alternative when medication can no longer provide reliable efficacy. It is a surgical therapy involving the implantation of one or more electrodes into specific regions of the brain, which deliver electrical stimulation to modulate or disrupt abnormal patterns of neural signaling [14]. To date, most applications of DBS for PD target the subthalamic nucleus (STN) and the internal globus pallidus (GPi) [15]. Over the last 20 years, STN/GPi DBS has become a well-established option for the treatment of PD patients with motor fluctuations and dyskinesias [15]. Under the condition that patients are responsive to L-DOPA treatment before surgery [16–18], DBS can reproduce the effects of L-DOPA and treat distal motor symptoms, i.e., tremor, rigidity, and akinesia [19, 20]. However, the effect on axial disturbances, i.e., gait and postural instability, remains controversial, sometimes even worsens gait [20–25].

1.4 Targeting of DBS

Before the DBS procedure, a neurosurgeon uses magnetic resonance imaging (MRI) and postoperative computed tomography (CT) to pinpoint the exact three-dimensional coordinates of the target area within the brain. Misplacement of the DBS electrode may lead to various adverse effects and unexpected outcomes [26]. Even a deviation from the intended target of 2 mm can result in unwanted side effects that negate the intended therapeutic response [27]. While the spatial resolution of modern MRI and CT scanners continue to improve, detailed nuclear anatomy remains impossible to discern. Since there are individual differences in anatomy, initial target selection is only approximate, and many experts agree that the final targeting should be performed using intraoperative microelectrode recordings (MER). MER is used to monitor the activity of neurons in the target area and thereby identify the precise brain target that will be stimulated. There are several reasons for using microelectrode recordings as an aid to localize boundaries of certain brain areas and surrounding structures in DBS: (1) Borders between white and gray matter are easily identified, as white matter is usually very quiet in comparison to gray matter. (2) Different nuclei have different patterns of activity which can serve to reliably distinguish among the nuclei. MER along preplanned trajectories are often used for improved delineation of the location of the STN during DBS surgery for PD [28].

In this review, we summarize the effect of DBS on gait and postural disturbances in PD. For this purpose, we first describe STN/GPi DBS and the two promising

DBS targets for gait and postural disturbances: pedunculopontine nucleus (PPN) and substantia nigra pars reticulata (SNr). In each section, we discuss the proposed mechanism(s) for the DBS target to treat gait and postural disturbances, a discussion of stimulation parameters (i.e., location, frequency, amplitude, and pulse width) used in existing studies, and the localization criteria for each target.

2 Methods

PubMed was the main database used to conduct the literature search. Keywords searched were as follows: Parkinson's disease, gait disorders, deep brain stimulation, freezing of gait, postural disturbances, fall, gait disorder, subthalamic nucleus, substantia nigra pars reticulata, and pedunculopontine nucleus. There was no specific time range used to limit the search. Meta-analysis was used for summarize the effects of STN/GPi DBS on gait, postural instability, and tremor.

3 STN/GPi DBS

Although postural instability and falls are thought to respond poorly to dopaminergic treatment [29], STN/GPi DBS in combination with L-DOPA treatment reduces postural instability by increasing motor abilities and specific postural related mechanisms, leading to a reduction in falls [16]. STN/GPi DBS itself has a controversial and less sustained effect on gait and postural disturbances, sometimes worsening these symptoms [20–25]. Recent long-term studies have reported a gradual decline in the effectiveness of STN/GPi DBS on gait disturbances and axial symptoms over 3 years [30, 31], 5 years [21, 32–34], 8 years [22], or 10 years [35, 36], in contrast to stable improvements of tremor rigidity and limb bradykinesia (Fig. 1).

3.1 Neuroanatomy of STN/GPi

The STN is subdivided into different territories, motor, oculomotor, associative, and limbic, each with different connections and functions. The large dorsolateral portion of the STN corresponds to the motor territory; the ventromedial portion to the associative territory and the medial tip to the limbic territory of the STN. Most STN neurons are glutamatergic projection neurons and provide a powerful excitatory input to the external segment of the GP (GPe) and to the two output structures of the basal ganglia, the GPi and SNr [37] (Fig. 2).

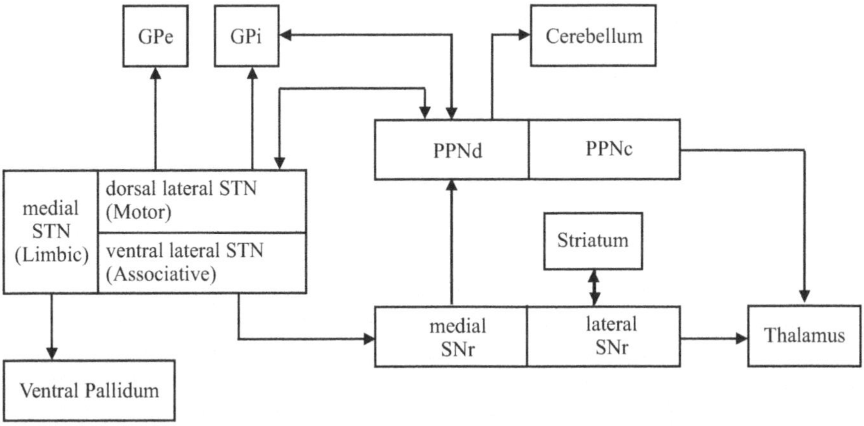

Fig. 1 Connectivity of DBS targets for gait and postural disturbances in Parkinson's disease. STN is subdivided into a large dorsolateral motor territory, a ventromedial associative territory, and a medial limbic territory. Each territory receives inputs from different areas of the cerebral cortex and provides output to different target nuclei, including the internal segment (GPi) and external segment (GPe) of the globus pallidus, substantia nigra pars reticulata (SNr), and ventral pallidum [37]; the PPNd receives glutamatergic input from the STN; GABAergic input from the GPi; GABAergic input from the SNr [38]; the PPNd projects to the STN, GPi, and cerebellum; the PPNc projects to the caudate/putamen and thalamus [38]; the lSNr receives input from the striatum and projects to thalamus; the mSNr receives input from the STN and projects to the PPN [39]

3.2 Proposed Mechanisms of Action

Stimulation of the STN may alleviate some potential aberrant suppression of PPN activity and/or compensate for neuron loss in the PPN, so as to improve gait and postural disturbances in PD [40]. PD patients with FOG showed a disruption between the STN/GPi and cortical regions, suggesting that FOG may be due to an overwhelming increase in the inhibitory output of the basal ganglia [39]. Stimulating the STN does not directly produce striatal dopamine release but may boost the dopamine motor system by inhibiting overexcited STN neurons [41, 42], reducing the neuronal synchronization in the vicinity of STN [43] and altering connectivity of thalamocortical pathways [44].

3.3 Effects of Stimulation Location and Frequency

Due to the anatomic connectivity differences between dorsal and ventral STN [45], DBS should target the dorsal STN, which has connections with sensorimotor brain regions [46]. Bilateral LFS (<80 Hz) [47] and HFS (>100 Hz) [48] have shown improvements in gait and postural disturbance when stimulated at more dorsal contacts compared to ventral STN. Others however did not find improvements in

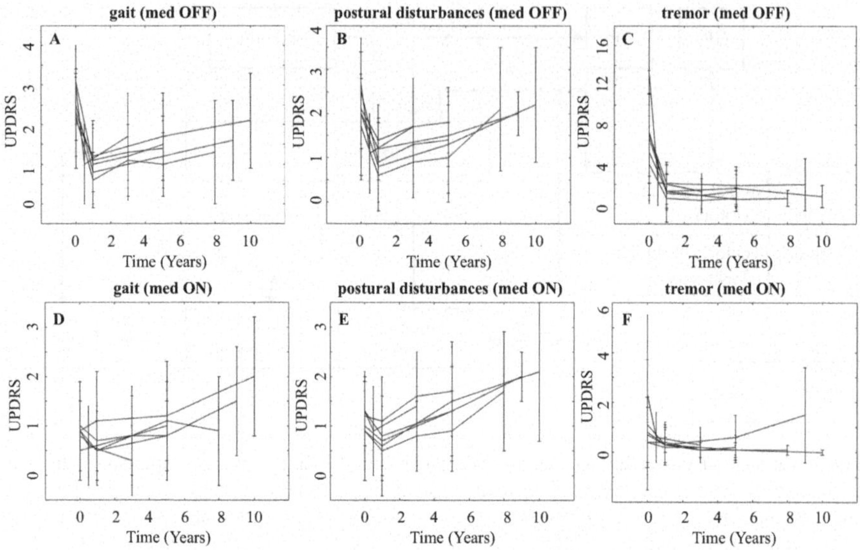

Fig. 2 Unified Parkinson's disease rating scale (UPDRS) of gait, postural instability, and tremor over time. Meta-analysis of results from clinical studies investigating the effects of the STN on gait (UPDRS III, item 29 and 30), postural instability (UPDRS III, item 28), and tremor (UPDRS III, item 20 and 21) over time [55–63]. Each line represents data reported from a single study. Note initial improvements in gait and postural instability at 6 months and 12 months followed by a progressive decline at later time points. All data are mean ± SD (UPDRS = 0, normal; UPDRS = 1, slight; UPDRS = 3, moderate; and UPDRS = 4, severe [12])

gait and balance with unilateral dorsal STN compared to unilateral ventral STN, but some cognitive function improvement [23]. It might be because the motor connections are diffusely distributed throughout the STN instead of sharply divided [49]. Ventral GPi DBS improve LID but not gait and akinesia, while dorsal GPi DBS improves akinesia but not LID [50]. When using intermediate contact, GPi DBS improved both LID and gait disorders [51]. Bilateral GPi DBS has little to no effect on postural control but significantly increases stride length, step length, and velocity during active walking, but less than STN DBS [16, 51]. Bilateral GPi DBS has better results than unilateral GPi DBS [19]. Some research reported a decrease number of FOG episodes at low frequencies (60–80 Hz) but not in gait performance [22, 52–54]; some reported immediate improvements in gait but the improvements were not sustained [55]. A recent study used 80 Hz STN DBS on one patient, and a reduction in falls and improvement of postural symptoms were observed at 1 year after surgery [56]. One quantitative study showed that LFS products similar or slightly better results compared to HFS for postural control and gait but not statistically significant [53]. However, there was no improvement in axial symptoms from the LFS after more than 3 years of HFS [57].

3.4 Targeting of STN/GPi DBS

Both MRI and MER are commonly used to localize STN/GPi during DBS surgery [58]. A 43% of GPi and 45% of STN mismatching of more than 3 mm using MRI per se between the theoretic coordinated and the final site of electrode location were reported [59]. Thus the proper electrode positioning of the STN/GPi may be improved by fine electrophysiological assessment [59, 60]. MERs of STN showed mean firing rate of 37 ± 17 Hz and an irregular firing pattern [61]. Another study showed 42.3 ± 22 Hz and an irregular or bursty discharge pattern [28]. The detection of the dorsolateral STN is based on clear-cut changes in electrical activity in the form of a sharp rise in the total power of the MERs [62], the tremor frequency (5 Hz) [63], and the beta oscillation (13–30 Hz) [64].

3.5 Limitations

(1) Currently, the therapeutic treatment of PD by DBS is an open-loop system where continuous stimulation is applied to the STN/GPi. Many of the existing shortcomings of DBS come from the open-loop nature of the implanted devices: The limited battery life requires multiple replacements. The vast majority of commercially available DBS devices are non-rechargeable and invasive surgery is required to replace the implantable pulse generator whenever the battery runs out. The battery's lifespan is determined primarily by how often the device is used and at what level of stimulation. The average battery life is <4 years [65]. If the stimulation is continuous, sometimes it does not achieve the maximum benefits and may trigger some side effects such as dyskinesia, speech problems, or depression. (2) The emergence of stimulation-related adverse effects is another important limitation of DBS therapy, occurring regularly in up to 13% of patients [14]. (3) Previous studies show that STN/GPi HFS is effective for tremor, long term of continuous DBS worsens gait and postural disturbances [20, 21, 24]. Some studies showed that STN/GPi LFS (60 Hz), compared to HFS, has beneficial effects on improving freezing of gait and other axial symptoms [66]. However, it is impossible to send HFS and LFS at the same time with current continuous DBS. Thus, to achieve both distal and axial symptoms, either another target needs to be involved or close-loop DBS needs to be developed for HFS and LFS separation.

4 PPN DBS

4.1 Neuroanatomy of PPN

The PPN, along with cuneiform nucleus (CN), comprises the mesencephalic locomotor region (MLR). MLR is an area of the brain stem that controls locomotion [67]. The PPN can be divided into the pars compacta (PPNc) and the pars dissipata (PPNd). The PPNc comprises mainly cholinergic neurons, whereas the PPNc contains both cholinergic and non-cholinergic cells. Subcortical afferents arise mainly from the GPi and SNr; efferents from cholinergic PPN neurons reach thalamic regions; projections from the PPNd provide innervation to the STN, SNc, and GPi [40] (Fig. 1).

4.2 Proposed Mechanisms of Action

Connections with the primary motor cortex, basal ganglia, thalamus, cerebellum and spinal cord play important roles in the regulation of movement by PPN region [68]. Reduced cerebellar connectivity pre-stimulation was partially restored after PPN DBS [69]. Gait, postural disturbances and falls are correlated with the loss of cholinergic neurons in the PPN [70–73]. PD monkey models with cholinergic lesion in the PPN showed resistance to L-DOPA treatment [73, 74]. The PPN is underactive in PD due to a combination of the degeneration and excessive inhibition. LFS PPN DBS could enhance PPN network activity, perhaps via disinhibition [75].

4.3 Effects of Stimulation Location and Frequency

In patients with advanced PD, bilateral LFS (25 Hz) PPN DBS associated with standard STN DBS may be useful in improving gait, yet PPN DBS alone is not as effective [76]. In this case, electrodes were implanted lateral to the medial lemniscus, in a region that encompasses the PPN [76]. When electrodes are implanted in the PPN, the cuneiform, and the subcuneiform nuclei, unilateral PPN DBS with 50–80 Hz is reported to mildly alleviate gait and falls but not significantly [77]. Bilateral PPN DBS at 25 Hz does improve FOG and frequency of falls at 1 year but no major effect in gait and posture [78]. Combination of caudal zona incerta (cZi)130 Hz with PPN <60 Hz postural stability became impaired whereas cZi stimulation <40 Hz worsened tremor, rigidity, and bradykinesia, when 60 Hz was chosen for combine PPN and cZi DBS, PD symptom was slightly improved [79]. When electrodes are localized bilaterally within the PPN, improvement was shown in frequency of fall, but not in gait, when combined with L-DOPA [80]. PPN LFS

DBS improved the hind limb and forelimb scores but no other gait measurement in unilateral lesioned PD model rat [81] (Table 1).

4.4 *Targeting of PPN DBS*

Targeting the PPN during DBS surgery is a controversial issue. Conventional MRI is often not suitable for a clear visualization of the region [38]. A few MER studies attributes may help targeting the PPN: (1) mean firing rate of the PPN was 23.2 ± 15.6 Hz; (2) spontaneous neuronal firing rate and burst discharge rate were significantly different between the dorsal PPN and in the PPN; and (3) theta and beta band oscillation were present in PPN LFPs [86, 87]. Within the PPN regions, 57% of the neurons fired randomly while 21% of the neurons exhibited bursty firing [87]. But still no hallmarks of identification of PPN were found compared to STN identification.

4.5 *Limitations*

Stimulation amplitude determines the spatial extent of neural activation. The PPN is a small midbrain region in comparison to the spatial extension of the stimulation effect produced by the microelectrode [38]. The lack of clarity of the PPN contributes to the difficulty in targeting and determining the exact localization of the electrodes implanted in human subjects suffering from neurodegenerative disorders. It is likely that DBS in the PPN region affects the neighbor structures. In humans the PPN overlaps with the posterior part of the substantia nigra (SN), so that it is presumably impossible to constrain stimulation to the PPN without also altering the SN. Hence, the observed effects on discrimination performance may at least to some degree stem from a modulation of activity in the SN [88]. Because of the location, PPN DBS may be risker than other DBS surgeries [80]. Stimulation-related adverse events during LFS PPN DSB include paresthesia, pain, and temperature sensation; some patients develop oscillopsia (a visual disturbance) during LFS PPN DBS [77, 78]. In a recent study, two out of six patients developed several adverse effects [80].

5 SNr DBS

5.1 *Neuroanatomy of SNr*

A GABAergic projection from the medial SNr (mSNr) to the MLR controls locomotion, and another projection from the lateral SNr (lSNr) to the PPN controls

Table 1 Studies using PPN DBS and SNr DBS to treat gait and postural disturbances in Parkinson's disease

			Stimulation parameters						
DBS target	Bi–/unilateral	Subject	Frequency (Hz)	Amplitude	Pulse width (μs)	Targeting tech-nique(s)	Electrode design	Main results	Ref
PPN	Bilateral	Human ($n = 6$)	25	1.5–2 V	60	MRI/CT /MER	Medtronic 3389	Immediate 45% improvement in gait and posture items in the UPDRS then decline to a final 32% in the score after 3–6 months	[76]
PPN	Unilateral	Human ($n = 6$)	50–80	0.7–3.8 V	60	MRI/CT	Medtronic 3387	No improvement in the UPDRS gait and postural stability score after 3 or 12 months but significant reduction in FOG both at 3 and 12 months with continuous DBS	[34, 77]
PPN	Bilateral	Human ($n = 6$)	15–25	1.2–3.8 V	60	MRI/CT	Medtronic 3389	FOG was improved after 1 year; gait measurement was significantly improved in 1 patient, moderately improved in 4 patients, and worsened in 1 patient	[78]
PPN and cZi	Bilateral	Human (n = 4)	60	N/A	60	MRI/CT	Medtronic 3389	Combined PPN/cZi DBS significantly improved gait scores in the UPDRS compared to cZi DBS alone	[79]
PPN	Bilateral	Human ($n = 6$)	35	0–5 V	60	MRI/CT	Medtronic 3389	Two patients were excluded due to adverse events; FOG, but not in gait measurement, was improved in 3 patients	[80]

PPN	Unilateral	Rats ($n = 14$)	25	100–600 μA	60	Stereotaxic	Platinum-Iridium (Plexon Inc.)	Hind limb and impaired forelimb movement was improved but not in other gait measurements	[115]
SNr	Bilateral	Human ($n = 7$)	159	1.4–3.5 V	60	MRI/CT	Medtronic 3389	Axial motor symptoms and braking capacity of fall was improved, but not distal motor symptoms	[82]
SNr and STN	Bilateral	Human ($n = 12$)	125	2.7 V	60	MRI/CT	Medtronic 3389	FOG was improved, but no global effect on axial motor impairment, after 3 weeks	[83]
SNr	Bilateral	Human ($n = 12$)	130	0.3–3.5 V	60/90	MRI/CT	Medtronic 3389	SNr DBS regulated temporal parameters of gait (swing time asymmetry)	[94]
SNr	Bilateral	Cats ($n = 27$)	50–100	30 μA	0.2 ms	Stereotaxic	Stainless Steel (Plastics One, Inc.)	Lateral SNr DBS attenuated and blocked muscle-tone suppression; medial SNr DBS reduced the number of step cycles and disturbed the rhythmic alternation of limb movement	[84]
SNr	Unilateral	Rats ($n = 6$)	130	50–200 μA	90	Stereotaxic	Platinum-Iridium (Micro-Probes)	Medial SNr DBS, but not lateral SNr DBS, improved rat's ability to maintain walking speed	[85]

postural muscle tone [89]. It has been demonstrated in a neuroanatomical study that the medial PPN, which approximately corresponds to the MLR, receives afferents from the mSNr, whereas the lateral PPN, which corresponds to the inhibitory region, receives afferents from the lSNr [90] (Fig. 1).

5.2 *Proposed Mechanisms of Action*

HFS SNr DBS could potentially restore the normal function of the brain stem centers that control muscle tone and locomotion and thereby alleviate gait and postural disturbances in PD [84]. Also, it is plausible that HFS SNr affects STN neuronal activity from current spread. It is also possible that SNr DBS may have increased SNc neuronal activity, since SNc is dorsal to and interdigitated among SNr. So that the excitation of SNc axons could have increased dopaminergic release in the striatum [91, 92].

5.3 *Effects of Stimulation Location and Frequency*

Medial SNr, not lateral SNr, GABA antagonist injection induced strong behavioral activation, turning, and dystonia-like neck and body postures [93]. Bilateral HFS medial SNr DBS dramatically improves axial Parkinsonian motor symptoms (gait and balance disorders) and braking capacity of fall but has no effect on distal Parkinsonian motor symptoms (segmental akinesia, rigidity, and tremor) [82]. Compared to standard STN DBS, combination of STN and SNr DBS with HFS showed slight improvement of FOG, but not postural instability. This improvement only happened immediately but not after 3 weeks [83]. In this case, the lateral SNr was implanted. Another study showed that stimulation of the medial SNr could modulate the temporal integration of gait, as opposed to the lateral part of the SNr [94]. In cats, local inhibition of SNr induces locomotion and reversed during HFS [84, 89]. Also, mSNr DBS improves the rat's ability to maintain walking speed but not lSNr and is associated with a decrease in ipsilateral circling in 6-OHDA rat model [85] (Table 1). SNr lesion caused dramatic changes in postural but had no effect on impaired reaching and hemineglect in hemiparkinsonian marmoset [95]. HFS on lateral SNr DBS at 130 Hz improved forelimb akinesia in a rat model of PD, but not LFS [92].SNr-HFS may have resulted from the attenuation of SNr neuronal spiking activity through a depolarization block phenomenon [92]. Unlike PPN DBS, concomitant stimulation of the SNr and STN is safe and well-tolerated [83].

5.4 Targeting of SNr DBS

MRI showed nonlinear changes of SNr patterns of during PD progression, from higher iron status at the beginning of PD to a dynamic decrease as disease advances [96]. So far, studies about SNr MERs are very limited in either human or rat, and most of them did not mention the subregions of SNr, namely, the mSNr and lSNr. In clinical studies, the anatomical closeness of SNr and STN enables the ventral contacts of electrode occasionally positioned in SNr when conducting current STN DBS [97] (Fig. 3). The firing rates in PD patients during DBS surgeries varied from 71 ± 23 spikes/s to 30 ± 13 spikes, and amplitude was reported as 46.7 ± 31.1 μV [28, 61, 98], whereas in healthy rat, the SNr was reported with 19.4 ± 1.2 spikes/s ($n = 48$), and the firing pattern was 71% of the SNr neurons which had a tonic discharge, 29% fired randomly, and 0% showed a bursting firing pattern [40]. In another report, the firing rate ranged from 10.19 to 41.16 spikes/s with a mean of 22.49 ± 0.92 spikes/s ($n = 89$), and the firing pattern was 56.18% regular, 37.08% irregular, and 6.74% bursting neurons [99]. One study that was specifically recorded from the subregions of SNr but only the GABAergic neurons reported that the mean firing rate was 21.98 ± 1.27 spikes/s ($n = 50$) in the lSNr and 23.15 ± 1.33 spikes/s ($n = 39$) in the mSNr. In normal rats, for neurons recorded within the lSNr, the percentage of regular firing neurons was 54.00%, while 40.00% of the neurons had an irregular firing and 6.00% of neurons fired in burst. Upon evaluation of neurons recorded within the mSNr, the firing pattern distribution was 58.97%

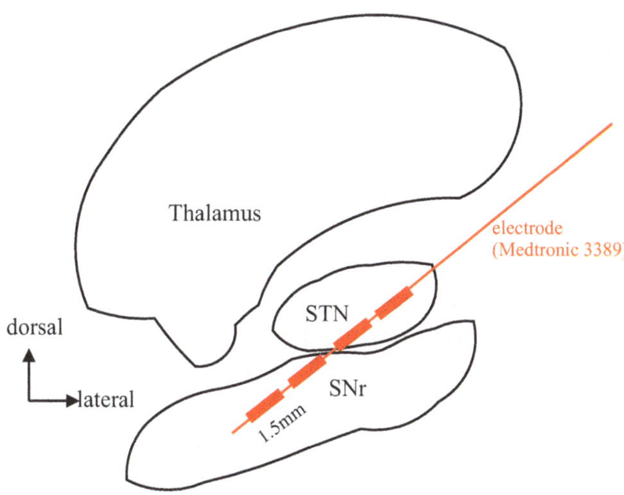

Fig. 3 Example of single FDA-approved DBS electrode design to apply co-stimulation of STN and medial SNr to treat the distal and axial symptoms of Parkinson's disease. Ventral stimulating contacts of the DBS electrode (Medtronic 3389) could be used to treat the gait and postural disturbances in PD simultaneously with dorsal stimulating contacts in the STN to treat tremor, bradykinesia, and rigidity

regular, 33.33% irregular, and 7.70% bursting neurons. In this study, no significant difference was found between the mSNr and lSNr in normal rat model [100]. An advanced targeting technique called Stimulus Pulse Aligned Coherence analysis in Evoked Recordings (SPACER) showed promising results to distinguish the mSNr and lSNr. SPACER is a technique that analyze the coherence of two evoked LFPs at adjacent locations, using a 0.5 Hz of stimulation and stimulus time alignment. It revealed that the mSNr and lSNr have an opposite pattern of coherence between LFPs in adjacent locations [101].

5.5 Limitations

There is a paucity of clinical and pre-clinical studies of SNr DBS. Larger number of groups are needed for further validation. Although no side effects of SNr DBS have been reported, more studies are needed to confirm the safety of this approach.

6 Discussion

Gait and postural disturbances in PD are responsible for high morbidity and mortality. Current STN/GPi DBS has controversial effects on gait and postural disturbances, and long-term studies showed a gradual decline in its effectiveness. Pilot studies of LFS PPN DBS and HFS SNr DBS have showed promising results. On the one hand, PPN DBS carries higher surgical risk of adverse effects than other DBS targets. On the other hand, SNr DBS appears to be safe and well-tolerated, though more studies targeting the medial sites remain unexplored in humans. Besides the close proximity of SNr and STN, ventral stimulating contacts of current STN DBS electrodes are easily positioned in the SNr, which increases the practicality of this approach.

This review highlights the critical importance of targeting for DBS treatment of gait and postural disturbances in PD. MRI and MER are typically combined to localize DBS target, as misplacement of DBS electrodes may lead to unexpected outcomes. MRI of STN caused 45% of mismatching of more than 3 mm between the theoretic coordinated and the final site of electrodes [59]. MRI of the PPN is not suitable for clear visualization [38]. MER may therefore enable more precise targeting than MRI alone. MER of STN/GPi is well characterized, whereas PPN MER does not show hallmarks suitable for identification. To our knowledge, no studies of SNr MERs exist.

Multiple studies support the assertion that DBS in subregions of a target will make a dramatic difference in improving the effectiveness of DBS to treat gait and postural disturbances. Future studies should consider the effects of DBS on subregions of targets due to the heterogeneity exists in the STN, PPN, and SNr. Sensorimotor regions of the dorsolateral STN DBS showed better improvement of gait and postural instability in PD [46]. Differences of neuron types and connections in the PPNc and PPNd may explain the controversial results of DBS effects of PPN DBS on axial symptoms. Connection with MLR of the mSNr may make this subregion a more suitable target to treat gait and postural instability. One key to investigate this effect are advancements in more precise localization of the DBS leads. Indeed, misplacement is a likely reason for the controversial results of PPN DBS and SNr DBS on gait and postural disturbances of PD. Clinical validation of techniques to target subregions, such as SPACER [101], is needed to improve the precision of DBS targeting. Also, future studies should report where within the target the electrode is implanted to aid in understanding more fully the DBS mechanisms of action.

The development of closed-loop DBS systems are expected to further improve the treatment of gait and postural disturbances in PD. There are parallels with cardiac pacemaker technology, which like DBS uses a neurostimulator to generate and deliver electric pulses into the deep brain areas through extension wires and electrodes. Cardiac stimulation devices have been in clinical use since 1963 [102]. Cardiac pacemakers currently can function to sense and respond to atrial activity in closed-loop mode. Efforts to bring similar concepts to DBS devices have been delayed in part due to the complexity of brain signals. Upon the detection of abnormalities during monitoring, closed-loop DBS applies responsive electrical stimulation. Over the past decade, it has become apparent that closed-loop DBS produces more clinical improvement and less undesirable side effects [103, 104] while in addition improving battery life and thus reducing the trauma of repetitive surgery [105, 106]. One promising approach to closed-loop DBS is to trigger stimulation pulses using the dynamic changes in the subcortical which are thought to lead to motor symptoms of PD. It appears that LFPs may provide the most relevant biomarker to close the loop in PD patient [75]. Also, it may be the case that a single biomarker is insufficient [107–109], so multiple biomarkers may be more effective for closed-loop control [110]. Searching for the relevant neural signatures that correlate with different PD symptoms, including biomarkers correlated to gait and postural disturbances in PD, is a top priority in this field. The development of DBS electrodes that can simultaneously stimulate and record makes this a feasible approach [111–115].

In summary, the need for better treatments of the gait and postural disturbances of PD has driven the field to identify suitable targets for DBS, with the primary emphasis on the STN, PPN, and SNr. An increasing awareness of the importance of the importance of DBS location should drive technological advancements to improve targeting of DBS electrodes. Accurate and precise targeting together with detailed reporting of DBS electrode location will further advance our understanding of the mechanisms of DBS in current targets and may aid in discovering new ones.

References

1. C. Marras, J.C. Beck, J.H. Bower, E. Roberts, B. Ritz, G.W. Ross, et al., Prevalence of Parkinson's disease across North America. NpjParkinson's Dis. **4**, 1–7 (2018). https://doi.org/10.1038/s41531-018-0058-0
2. S.L. Kowal, T.M. Dall, R. Chakrabarti, M.V. Storm, A. Jain, The current and projected economic burden of Parkinson's disease in the United States. Mov. Disord. **28**(3), 10–15 (2013). https://doi.org/10.1002/mds.25292
3. J.M. Fearnley, A.J. Lees, Ageing and Parkinson' disease: substantia nigra regional selectivity. Brain, 114, 2283–2301 (1991)
4. M. Politis, K. Wu, S. Molloy, P.G. Bain, K.R. Chaudhuri, P. Piccini, Parkinson's disease symptoms: the patient's perspective. Mov. Disord. **25**(11), 1646–1651 (2010). https://doi.org/10.1002/mds.23135
5. N. Giladi, T.A. Treves, E.S. Simon, H. Shabtai, Y. Orlov, B. Kandinov, et al., Freezing of gait in patients with advanced Parkinson's disease. J. Neural Transm. (Vienna) **108**, 53–61 (2001)
6. J.G. Nutt, B.R. Bloem, N. Giladi, M. Hallett, F.B. Horak, A. Nieuwboer, Freezing of gait: moving forward on a mysterious clinical phenomenon. Lancet Neurol. **10**(8), 734–744 (2011). https://doi.org/10.1016/S1474-4422(11)70143-0
7. B.R. Bloem, J.M. Hausdorff, J.E. Visser, N. Giladi, Falls and freezing of gait in Parkinson's disease: a review of two interconnected, episodic phenomena. Mov. Disord. **19**(8), 871–884 (2004). https://doi.org/10.1002/mds.20115
8. J.J. Crouse, J.R. Phillips, M. Jahanshahi, A.A. Moustafa, Postural instability and falls in Parkinson's disease. Rev. Neurosci. **27**(5), 549–555 (2016). https://doi.org/10.1515/revneuro-2016-0002
9. G.K. Wenning, G. Ebersbach, M. Verny, K.R. Chaudhuri, K. Jellinger, A. McKee, et al., Progression of falls in postmortem-confirmed parkinsonian disorders. Mov. Disord. **14**(6), 947–950 (1999). https://doi.org/10.1002/1531-8257(199911)14:6<947::AID-MDS1006>3.0.CO;2-O
10. T. Masud, R.O. Morris, Epidemiology of falls. Age Ageing **30**, 3–7 (2001). https://doi.org/10.1093/ageing/30.suppl_3.3
11. M.W. Creaby, M.H. Cole, Gait characteristics and falls in Parkinson's disease: a systematic review and meta-analysis. Parkinsonism Relat. Disord. **57**, 1–8 (2018). https://doi.org/10.1016/j.parkreldis.2018.07.008
12. C.G. Goetz, B.C. Tilley, S.R. Shaftman, G.T. Stebbins, S. Fahn, P. Martinez-martin, et al., Movement disorder society-sponsored revision of the unified Parkinson's disease rating scale (MDS-UPDRS): scale presentation and clinimetric testing results. Mov. Disord. **23**(15), 2129–2170 (2008). https://doi.org/10.1002/mds.22340
13. G. Porras, P. De Deurwaerdere, Q. Li, M. Marti, R. Morgenstern, R. Sohr, et al., L-dopa-induced dyskinesia: beyond an excessive dopamine tone in the striatum. Sci. Rep. **4**, 3730 (2014). https://doi.org/10.1038/srep03730
14. P. Hickey, M. Stacy, Deep brain stimulation: a paradigm shifting approach to treat Parkinson's disease. Front. Neurosci. **10**, 173 (2016). https://doi.org/10.3389/fnins.2016.00173
15. A. Ramirez-Zamora, J.L. Ostrem, Globus pallidus interna or subthalamic nucleus deep brain stimulation for Parkinson disease a review. JAMA Neurol. **75**(3), 367–372 (2018). https://doi.org/10.1001/jamaneurol.2017.4321
16. P.P. Perrin, Bilateral subthalamic nucleus stimulation improves balance control in Parkinson's disease. J. Neurol. Neurosurg. Psychiatry **76**, 780–788 (2005). https://doi.org/10.1136/jnnp.2004.047829
17. M. Pötter-Nerger, J. Volkmann, Deep brain stimulation for gait and postural symptoms in Parkinson's disease. Mov. Disord. **28**(11), 1609–1615 (2013). https://doi.org/10.1002/mds.25677
18. S. Vercruysse, W. Vandenberghe, L. Münks, B. Nuttin, H. Devos, A. Nieuwboer, Effects of deep brain stimulation of the subthalamic nucleus on freezing of gait in Parkinson's disease:

a prospective controlled study. J. Neurol. Neurosurg. Psychiatry **85**, 872–878 (2014). https://doi.org/10.1136/jnnp-2013-306336

19. P. Krack, P. Pollak, P. Limousin, D. Hoffmann, J. Xie, A. Benazzouz, A.L. Benabid, Subthalamic nucleus or internal pallidal stimulation in young onset Parkinson's disease. Brain **121**(Pt 3), 451–457 (1998). https://doi.org/10.1093/brain/121.3.451
20. M.C. Rodriguez-Oroz, J.A. Obeso, A.E. Lang, J.L. Houeto, P. Pollak, S. Rehncrona, et al., Bilateral deep brain stimulation in Parkinson's disease: a multicentre study with 4 years follow-up. Brain **128**(Pt 10), 2240–2249 (2005). https://doi.org/10.1093/brain/awh571
21. S. Chabardes, V. Fraix, C. Ardouin, A. Koudsie, P.D. Limousin, D. Ph, Five-year follow-up of bilateral stimulation of the subthalamic nucleus in advanced parkinson's disease. N. Engl. J. Med. **349**(20), 1925–1934 (2003)
22. A. Fasano, J. Herzog, E. Seifert, H. Stolze, D. Falk, J. Volkmann, Modulation of gait coordination by subthalamic stimulation improves freezing of gait. Mov. Disord. **26**(5), 844–851 (2011). https://doi.org/10.1002/mds.23583
23. M.E. McNeely, T. Hershey, M.C. Campbell, S.D. Tabbal, M. Karimi, J.M. Hartlein, et al., Effects of deep brain stimulation of dorsal versus ventral subthalamic nucleus regions on gait and balance in Parkinson's disease. J. Neurol. Neurosurg. Psychiatry **82**(11), 1250–1255 (2011). https://doi.org/10.1136/jnnp.2010.232900
24. W.M.M. Schüpbach, N. Chastan, M.L. Welter, J.L. Houeto, V. Mesnage, A.M. Bonnet, et al., Stimulation of the subthalamic nucleus in Parkinson's disease: a 5 year follow up. J. Neurol. Neurosurg. Psychiatry **76**(12), 1640–1644 (2005). https://doi.org/10.1136/jnnp.2005.063206
25. B.F.L. van Nuenen, R.A.J. Esselink, M. Munneke, J.D. Speelman, T. van Laar, B.R. Bloem, Postoperative gait deterioration after bilateral subthalamic nucleus stimulation in Parkinson's disease. Mov. Disord. **23**(16), 2404–2406 (2008). https://doi.org/10.1002/mds.21986
26. D.T.M. Chan, X.L. Zhu, J.H.M. Yeung, V.C.T. Mok, E. Wong, C. Lau, et al., Complications of deep brain stimulation: a collective review. Asian J. Surg. **32**(4), 258–263 (2009). https://doi.org/10.1016/S1015-9584(09)60404-8
27. E.B. Montgomery, Microelectrode targeting of the subthalamic nucleus for deep brain stimulation surgery. Mov. Disord. **27**(11), 1387–1391 (2012). https://doi.org/10.1002/mds.25000
28. A. Benazzouz, S. Breit, A. Koudsie, P. Pollak, Intraoperative microrecordings of the subthalamic nucleus in Parkinson's disease. Mov. Disord. **17**, 145–149 (2002). https://doi.org/10.1002/mds.10156
29. A.M. Bonnet, Y. Loria, M.H. Saint-Hilaire, F. Lhermitte, Y. Agid, Does long-term aggravation of Parkinson's disease result from nondopaminergic lesions? Neurology **37**(9), 1539–1542 (1987). https://doi.org/10.1212/WNL.37.9.1539
30. J. Volkmann, A. Albanese, J. Kulisevsky, A. Tornqvist, J. Houeto, B. Pidoux, et al., Long-term effects of pallidal or subthalamic deep brain stimulation on quality of life in Parkinson's disease. Mov. Disord. **24**(8), 1154–1161 (2009). https://doi.org/10.1002/mds.22496
31. F.M. Weaver, K.A. Follett, M. Stern, C.L. Harris, J. Rothlind, E.C. Lai, ... R. Simpson, Randomized trial of deep brain stimulation for Parkinson disease (2012)
32. A. Kishore, R. Rao, S. Krishnan, D. Panikar, G. Sarma, M.P. Sivasanakaran, S. Sarma, Long-term stability of effects of subthalamic stimulation in Parkinson's disease: Indian experience. Mov. Disord. **25**(14), 2438–2444 (2010). https://doi.org/10.1002/mds.23269
33. A. Merola, M. Zibetti, S. Angrisano, L. Rizzi, V. Ricchi, C.A. Artusi, et al., Parkinson's disease progression at 30 years: a study of subthalamic deep brain-stimulated patients. Brain J. Neurol. **134**(Pt 7), 2074–2084 (2011). https://doi.org/10.1093/brain/awr121
34. E. Moro, A.M. Lozano, P. Pollak, Y. Agid, S. Rehncrona, J. Volkmann, et al., Long-term results of a multicenter study on subthalamic and pallidal stimulation in Parkinson's disease. Mov. Disord. **25**(5), 578–586 (2010b). https://doi.org/10.1002/mds.22735
35. A. Castrioto, A.M. Lozano, Y.Y. Poon, A.E. Lang, M. Fallis, E. Moro, Ten-year outcome of subthalamic stimulation in Parkinson disease: a blinded evaluation. Arch. Neurol. **68**(12), 1550–1556 (2011). https://doi.org/10.1001/archneurol.2011.182

36. M. Zibetti, A. Merola, L. Rizzi, V. Ricchi, S. Angrisano, Beyond nine years of continuous subthalamic nucleus deep brain stimulation in Parkinson's disease. Mov. Disord. **26**(13), 2327–2334 (2011). https://doi.org/10.1002/mds.23903
37. E.E. Benarroch, Implications of subthalamic nucleus and its connections stimulation, 1991–1996 (2008)
38. C. Hamani, E. Moro, A.M. Lozano, The pedunculopontine nucleus as a target for deep brain stimulation. J. Neural Transm. **118**, 1461–1468 (2011). https://doi.org/10.1007/s00702-010-0547-8
39. J.M. Shine, E. Matar, P.B. Ward, S.J. Bolitho, M. Gilat, M. Pearson, et al., Exploring the cortical and subcortical functional magnetic resonance imaging changes associated with freezing in Parkinson's disease. Brain **136**, 1204–1215 (2018). https://doi.org/10.1093/brain/awt049
40. P.A. Pahapill, A.M. Lozano, The pedunculopontine nucleus and Parkinson's disease, 1767–1783 (2000)
41. A. Abosch, S. Kapur, A.E. Lang, D. Hussey, E. Sime, J. Miyasaki, et al., Stimulation of the subthalamic nucleus in Parkinson's disease does not produce striatal dopamine release. Neurosurgery **53**(5), 1095–1102 (2003). https://doi.org/10.1227/01.NEU.0000088662.69419.1B
42. C.H. Tai, M.K. Pan, J.J. Lin, C.S. Huang, Y.C. Yang, C.C. Kuo, Subthalamic discharges as a causal determinant of parkinsonian motor deficits. Ann. Neurol. **72**(3), 464–476 (2012). https://doi.org/10.1002/ana.23618
43. A. Eusebio, W. Thevathasan, L. Doyle Gaynor, A. Pogosyan, E. Bye, T. Foltynie, et al., Deep brain stimulation can suppress pathological synchronisation in parkinsonian patients. J. Neurol. Neurosurg. Psychiatry **82**(5), 569–573 (2011). https://doi.org/10.1136/jnnp.2010.217489
44. J. Kahan, L. Mancini, M. Urner, K. Friston, M. Hariz, E. Holl, et al., Therapeutic subthalamic nucleus deep brain stimulation reverses cortico-thalamic coupling during voluntary movements in Parkinson's disease. PLoS One **7**(12), e50270 (2012). https://doi.org/10.1371/journal.pone.0050270
45. A. Parent, L.N. Hazrati, Functional anatomy of the basal ganglia. II. The place of subthalamic nucleus and external pallidium in basal ganglia circuitry. Brain Res. Rev. **20**(1), 128–154 (1995). https://doi.org/10.1016/0165-0173(94)00008-D
46. B.R. Aravamuthan, K.A. Muthusamy, J.F. Stein, T.Z. Aziz, H. Johansen-Berg, Topography of cortical and subcortical connections of the human pedunculopontine and subthalamic nuclei. NeuroImage **37**(3), 694–705 (2007). https://doi.org/10.1016/j.neuroimage.2007.05.050
47. H.M. Khoo, H. Kishima, K. Hosomi, T. Maruo, N. Tani, S. Oshino, et al., Low-frequency subthalamic nucleus stimulation in Parkinson's disease: a randomized clinical trial. Mov. Disord. **29**(2), 270–274 (2014). https://doi.org/10.1002/mds.25810
48. E.L. Johnsen, N. Sunde, P.H. Mogensen, K. Østergaard, MRI verified STN stimulation site – gait improvement and clinical outcome. Eur. J. Neurol. **18**, 746–753 (2010). https://doi.org/10.1111/j.1468-1331.2010.02962.x
49. L. Mallet, M. Schupbach, K. N'Diaye, P. Remy, E. Bardinet, V. Czernecki, et al., Stimulation of subterritories of the subthalamic nucleus reveals its role in the integration of the emotional and motor aspects of behavior. Proc. Natl. Acad. Sci. **104**(25), 10661–10666 (2007). https://doi.org/10.1073/pnas.0610849104
50. B. Bejjani, P. Damier, I. Arnulf, A.M. Bonnet, M. Vidailhet, D. Dormont, et al., Pallidal stimulation for Parkinson's disease: two targets? Neurology **49**(6), 1564–1569 (1997). https://doi.org/10.1212/WNL.49.6.1564
51. P. Krystkowiak, J.D. Guieu, Chronic bilateral pallidal stimulation and levodopa do not improve gait in the same way in Parkinson's disease: a study using a video motion analysis system. J. Neurol. **248**(11), 944–949 (2001)
52. C. Moreau, STN-DBS frequency effects on freezing of gait in advanced Parkinson disease (2008)

53. S. Vallabhajosula, I.U. Haq, N. Hwynn, G. Oyama, M. Okun, M.D. Tillman, C.J. Hass, Brain stimulation low-frequency versus high-frequency subthalamic nucleus deep brain stimulation on postural control and gait in Parkinson's disease: a quantitative study. Brain Stimul. **8**(1), 64–75 (2015). https://doi.org/10.1016/j.brs.2014.10.011
54. T. Xie, U.J. Kang, P. Warnke, Effect of stimulation frequency on immediate freezing of gait in newly activated STN DBS in Parkinson's disease. J. Neurol. Neurosurg. Psychiatry **83**(10), 1015–1017 (2012). https://doi.org/10.1136/jnnp-2011-302091
55. V. Ricchi, M. Zibetti, S. Angrisano, A. Merola, N. Arduino, C.A. Artusi, et al., Transient effects of 80 Hz stimulation on gait in STN DBS treated PD patients: a 15 months follow-up study. Brain Stimul. **5**(3), 388–392 (2012). https://doi.org/10.1016/j.brs.2011.07.001
56. M.D. Mendonça, R. Barbosa, A. Seromenho-Santos, C. Reizinho, P. Bugalho, R. Miguel, et al., Early use of 80 Hz subthalamic stimulation in Parkinson's disease as an alternative for High-frequency stimulation induced gait changes and postural instability. Brain Stimul. **11**(3), 620–622 (2018). https://doi.org/10.1016/j.brs.2017.12.005
57. C. Sidiropoulos, R. Walsh, C. Meaney, Y.Y. Poon, M. Fallis, E. Moro, Low-frequency subthalamic nucleus deep brain stimulation for axial symptoms in advanced Parkinson's disease. J. Neurol. **260**(9), 2306–2311 (2013). https://doi.org/10.1007/s00415-013-6983-2
58. N.A. Hamid, R.D. Mitchell, P. Mocroft, G.W.M. Westby, J. Milner, H. Pall, Targeting the subthalamic nucleus for deep brain stimulation: technical approach and fusion of pre- and postoperative MR images to define accuracy of lead placement. J. Neurol. Neurosurg. Psychiatry **76**(3), 409–414 (2005). https://doi.org/10.1136/jnnp.2003.032029
59. J. Guridi, M.C. Rodriguez-Oroz, A.M. Lozano, E. Moro, A. Albanese, B. Nuttin, et al., Targeting the basal ganglia for deep brain stimulation in Parkinson's disease. Neurology **55**(12 Suppl 6), S21–S28 (2000)
60. E. Cuny, D. Guehl, P. Burbaud, C. Gross, V. Dousset, A. Rougier, Lack of agreement between direct magnetic resonance imaging and statistical determination of a subthalamic target: the role of electrophysiological guidance. J. Neurosurg. **97**(3), 591–597 (2002). https://doi.org/10.3171/jns.2002.97.3.0591
61. W.D. Hutchison, R.J. Allan, H. Opitz, R. Levy, J.O. Dostrovsky, A.E. Lang, A.M. Lozano, Neurophysiological identification of the subthalamic nucleus in surgery for Parkinson's disease. Ann. Neurol. **44**(4), 622–628 (1998). https://doi.org/10.1002/ana.410440407
62. A. Moran, I. Bar-Gad, H. Bergman, Z. Israel, Real-time refinement of subthalamic nucleus targeting using Bayesian decision-making on the root mean square measure. Mov. Disord. **21**(9), 1425–1431 (2006). https://doi.org/10.1002/mds.20995
63. H.J. Lee, W.W. Lee, S.K. Kim, H. Park, H.S. Jeon, H.B. Kim, et al., Tremor frequency characteristics in Parkinson's disease under resting-state and stress-state conditions. J. Neurol. Sci. **362**, 272–277 (2016). https://doi.org/10.1016/j.jns.2016.01.058
64. A. Zaidel, A. Spivak, B. Grieb, H. Bergman, Z. Israel, Subthalamic span of β oscillations predicts deep brain stimulation efficacy for patients with Parkinson's disease. Brain **133**(Pt 7), 2007–2021 (2010). https://doi.org/10.1093/brain/awq144
65. M. Bin-Mahfoodh, C. Hamani, E. Sime, A.M. Lozano, Longevity of batteries in internal pulse generators used for deep brain stimulation. Stereotact. Funct. Neurosurg. **80**(1–4), 56–60 (2003). https://doi.org/10.1159/000075161
66. T. Xie, M. Padmanaban, L. Bloom, E. MacCracken, B. Bertacchi, A. Dachman, P. Warnke, Effect of low versus high frequency stimulation on freezing of gait and other axial symptoms in Parkinson patients with bilateral STN DBS: a mini-review. Transl. Neurodegener. **6**, 13 (2017). https://doi.org/10.1186/s40035-017-0083-7
67. E. Garcia-Rill, The pedunculopontine nucleus. Prog. Neurobiol. **36**(5), 363–389 (1991)
68. K.A. Muthusamy, B.R. Aravamuthan, M.L. Kringelbach, N. Jenkinson, N.L. Voets, H. Johansen-Berg, et al., Connectivity of the human pedunculopontine nucleus region and diffusion tensor imaging in surgical targeting. J. Neurosurg. **107**(4), 814–820 (2007). https://doi.org/10.3171/JNS-07/10/0814

69. P.M. Schweder, C. Joint, P.C. Hansen, A.L. Green, G. Quaghebeur, T.Z. Aziz, Chronic pedunculopontine nucleus stimulation restores functional connectivity. Neuroreport **21**, 1065–1068 (2010). https://doi.org/10.1097/WNR.0b013e32833ce607
70. N.I. Bohnen, M.L.T.M. Müller, R.A. Koeppe, S.A. Studenski, M.A. Kilbourn, K.A. Frey, R.L. Albin, History of falls in Parkinson disease is associated with reduced cholinergic activity. Neurology **73**(20), 1670–1676 (2009). https://doi.org/10.1212/WNL.0b013e3181c1ded6
71. N.I. Bohnen, K.A. Frey, S. Studenski, V. Kotagal, R.A. Koeppe, P.J.H. Scott, et al., Gait speed in Parkinson disease correlates with cholinergic degeneration. Neurology **81**(18), 1611–1616 (2013). https://doi.org/10.1212/WNL.0b013e3182a9f558
72. E.C. Hirsch, A.M. Graybiel, C. Duyckaerts, F. Javoy-Agid, Neuronal loss in the pedunculopontine tegmental nucleus in Parkinson disease and in progressive supranuclear palsy. Proc. Natl. Acad. Sci. USA **84**(16), 5976–5980 (1987)
73. C. Karachi, D. Grabli, F.A. Bernard, D. Tandé, N. Wattiez, H. Belaid, et al., Cholinergic mesencephalic neurons are involved in gait and postural disorders in Parkinson disease. J. Clin. Investig. **120**(8), 2745–2754 (2010). https://doi.org/10.1172/JCI42642
74. D. Grabli, C. Karachi, E. Folgoas, M. Monfort, D. Tande, S. Clark, et al., Gait disorders in parkinsonian monkeys with pedunculopontine nucleus lesions: a tale of two systems. J. Neurosci. **33**(29), 11986–11993 (2013). https://doi.org/10.1523/JNEUROSCI.1568-13.2013
75. W. Thevathasan, E. Moro, What is the therapeutic mechanism of pedunculopontine nucleus stimulation in Parkinson's disease? Neurobiol. Dis. **128**, 67–74 (2018). https://doi.org/10.1016/j.nbd.2018.06.014
76. A. Stefani, A.M. Lozano, A. Peppe, P. Stanzione, S. Galati, D. Tropepi, et al., Bilateral deep brain stimulation of the pedunculopontine and subthalamic nuclei in severe Parkinson's disease. Brain **130**, 1596–1607 (2007). https://doi.org/10.1093/brain/awl346
77. E. Moro, C. Hamani, Y. Poon, T. Al-khairallah, O. Dostrovsky, W.D. Hutchison, A.M. Lozano, Unilateral pedunculopontine stimulation improves falls in Parkinson's disease. Brain **133**(Pt 1), 215–224 (2010a). https://doi.org/10.1093/brain/awp261
78. V. Fraix, L. Goetz, C. Ardouin, J. Yelnik, M.U. Ferraye, B. Debu, et al., Effects of pedunculopontine nucleus area stimulation on gait disorders in Parkinson's disease. Brain **133**(Pt 1), 205–214 (2010). https://doi.org/10.1093/brain/awp229
79. S. Khan, S.S. Gill, L. Mooney, S. Khan, Combined pedunculopontine-subthalamic stimulation in Parkinson disease. Neurology **78**(14), 1090–1095 (2012). https://doi.org/10.1212/WNL.0b013e31824e8e96
80. M. Welter, A. Demain, C. Ewenczyk, PPNa-DBS for gait and balance disorders in Parkinson's disease: a double-blind, randomised study. J. Neurol. **262**, 1515–1525 (2015). https://doi.org/10.1007/s00415-015-7744-1
81. Z.W. Zhang, Low-frequency stimulation of the pedunculopontine nucleus affects gait and the neurotransmitter level in the ventrolateral thalamic nucleus in 6-OHDA Parkinsonian rats. Elsevier (2015) https://doi.org/10.1016/j.neulet.2015.06.006
82. N. Chastan, G.W.M. Westby, J. Yelnik, E. Bardinet, M.C. Do, Y. Agid, M.L. Welter, Effects of nigral stimulation on locomotion and postural stability in patients with Parkinson's disease. Brain **132**(1), 172–184 (2009). https://doi.org/10.1093/brain/awn294
83. D. Weiss, M. Walach, C. Meisner, M. Fritz, M. Scholten, S. Breit, et al., Nigral stimulation for resistant axial motor impairment in Parkinson's disease? A randomized controlled trial. Brain **136**, 2098–2108 (2013). https://doi.org/10.1093/brain/awt122
84. K. Takakusaki, T. Habaguchi, J. Ohtinata-Sugimoto, K. Saitoh, T. Sakamoto, Basal ganglia efferents to the brainstem centers controlling postural muscle tone and locomotion: a new concept for understanding motor disorders in basal ganglia dysfunction. Neuroscience **119**, 293–308 (2003). https://doi.org/10.1016/S0306-4522(03)00095-2
85. G.C. McConnell, W.M. Grill, Stimulation location within the substantia nigra pars reticulata differentially modulates gait in hemiparkinsonian rats. In *Proceedings of the 6thInternational IEEE EMBS Conference on Neural Engineering*, San Diego, CA, (2013)

86. S.A. Shimamoto, P.S. Larson, J.L. Ostrem, G.A. Glass, R.S. Turner, P.A. Starr, Physiological identification of the human pedunculopontine nucleus. J. Neurol. Neurosurg. Psychiatry **81**, 80–86 (2010). https://doi.org/10.1136/jnnp.2009.179069
87. M. Weinberger, C. Hamani, W.D. Hutchison, E. Moro, A.M. Lozano, J.O. Dostrovsky, Pedunculopontine nucleus microelectrode recordings in movement disorder patients. Exp. Brain Res. **188**(2), 165–174 (2008). https://doi.org/10.1007/s00221-008-1349-1
88. H. Strumpf, T. Noesselt, M.A. Schoenfeld, J. Voges, P. Panther, J. Kaufmann, et al., Deep brain stimulation of the pedunculopontine tegmental nucleus (PPN) influences visual contrast sensitivity in human observers. PLoS One **11**(5), e0155206 (2016). https://doi.org/10.1371/journal.pone.0155206
89. K. Takakusaki, K. Saitoh, H. Harada, M. Kashiwayanagi, Role of basal ganglia – brainstem pathways in the control of motor behaviors. Neurosci. Res. **50**, 137–151 (2004). https://doi.org/10.1016/j.neures.2004.06.015
90. T. Moriizumi, Y. Nakamura, H. Tokuno, Y. Kitao, M. Kudo, Topographic projections from the basal ganglia to the nucleus tegmenti pedunculopontinus pars compacta of the cat with special reference to pallidal projections. Exp. Brain Res. **71**(2), 298–306 (1988). https://doi.org/10.1007/BF00247490
91. Y.H. Fu, Y. Yuan, G. Halliday, Z. Rusznák, C. Watson, G. Paxinos, A cytoarchitectonic and chemoarchitectonic analysis of the dopamine cell groups in the substantia nigra, ventral tegmental area, and retrorubral field in the mouse. Brain Struct. Funct. **217**(2), 591–612 (2012). https://doi.org/10.1007/s00429-011-0349-2
92. A.C. Sutton, W. Yu, M.E. Calos, A.B. Smith, A. Ramirez-zamora, E.S. Molho, et al., Deep brain stimulation of the substantia nigra pars reticulata improves forelimb akinesia in the hemiparkinsonian rat. J. Neurophysiol. **109**, 363–374 (2013). https://doi.org/10.1152/jn.00311.2012
93. T. Wichmann, M.A. Kliem, M.R. Delong, Antiparkinsonian and behavioral effects of inactivation of the substantia nigra pars reticulata in hemiparkinsonian primates. Exp. Neurol. **424**, 410–424 (2001). https://doi.org/10.1006/exnr.2000.7572
94. D. Weiss, Effects of subthalamic and nigral stimulation on gait kinematics in Parkinson's disease. Front. Neurol. **8**(October), 1–8 (2017). https://doi.org/10.3389/fneur.2017.00543
95. J.M. Henderson, D. Stanic, D. Tomas, J. Patch, M.K. Horne, D. Bourke, D.I. Finkelstein, Postural changes after lesions of the substantia nigra pars reticulata in hemiparkinsonian monkeys. Behav. Brain Res. **160**, 267–276 (2005). https://doi.org/10.1016/j.bbr.2004.12.008
96. G. Du, M.M. Lewis, C. Sica, L. He, J.R. Connor, L. Kong, et al., Distinct progression pattern of susceptibility MRI in the substantia nigra of Parkinson's patients. Mov. Disord. **33**(9), 1423–1431 (2018). https://doi.org/10.1002/mds.27318
97. C.R. Camalier, P.E. Konrad, C.E. Gill, C. Kao, M.R. Remple, H.M. Nasr, et al., Methods for surgical targeting of the. STN in early-stage Parkinson's disease **5**(March), 1–6 (2014). https://doi.org/10.3389/fneur.2014.00025
98. S. Mrakic-sposta, S. Marceglia, M. Egidi, G. Carrabba, P. Rampini, M. Locatelli, et al., Extracellular spike microrecordings from the subthalamic area in Parkinson's disease. J. Clin. Neurosci. **15**, 559–567 (2008). https://doi.org/10.1016/j.jocn.2007.02.091
99. S. Breit, A. Martin, L. Lessmann, D. Cerkez, T. Gasser, J.B. Schulz, Bilateral changes in neuronal activity of the basal ganglia in the unilateral 6-hydroxydopamine rat model. J. Neurosci. Sci. **86**(6), 1388–1396 (2008). https://doi.org/10.1002/jnr.21588
100. Y. Wang, Q. Jun, J. Liu, U. Ali, Z. Hua, Y. Ping, et al., Changes in firing rate and pattern of GABAergic neurons in subregions of the substantia nigra pars reticulata in rat models of Parkinson's disease. Brain Res. **1324**, 54–63 (2010). https://doi.org/10.1016/j.brainres.2010.02.008
101. H. Li, G.C. McConnell, Microstimulation evoked neuronal activity in the substantia nigra pars reticulata in anesthetized rats. Brain Stimul. **12**(2), e66–e68 (2019)
102. D.A. Nathan, S. Center, C.y. Wu, W. Keller, An implantable synchronous pacemaker for the long term correction of complete heart block. Am. J. Cardiol. **11**(3), 362–367 (1963). https://doi.org/10.1016/0002-9149(63)90130-9

103. B. Rosin, M. Slovik, R. Mitelman, M. Rivlin-Etzion, S.N. Haber, Z. Israel, et al., Closed-loop deep brain stimulation is superior in ameliorating parkinsonism. Neuron **72**, 370–384 (2011). https://doi.org/10.1016/j.neuron.2011.08.023
104. F.J. Santos, R.M. Costa, F. Tecuapetla, Stimulation on demand: closing the loop on deep brain stimulation. Neuron **72**, 197–198 (2011). https://doi.org/10.1016/j.neuron.2011.10.004
105. M.N. Gasson, S.Y. Wang, T.Z. Aziz, J.F. Stein, K. Warwick, Towards a demand driven deep-brain stimulator for the treatment of movement disorders. In *3rd IEE International Seminar on Medical Applications of Signal Processing* (2005), https://doi.org/10.1049/ic:20050336
106. D. Graupe, I. Basu, D. Tuninetti, P. Vannemreddy, K.V. Slavin, Adaptively controlling deep brain stimulation in essential tremor patient via surface electromyography. Neurol. Res. **32**(9), 899–904 (2010). https://doi.org/10.1179/016164110X12767786356354
107. M. Cassidy, P. Mazzone, A. Oliviero, A. Insola, P. Tonali, V. Di Lazzaro, P. Brown, Movement-related changes in synchronization in the human basal ganglia. Brain **125**(Pt 6), 1235–1246 (2002). https://doi.org/10.1093/brain/awf135
108. S. Little, A. Pogosyan, A.A. Kuhn, P. Brown, Beta band stability over time correlates with Parkinsonian rigidity and bradykinesia. Exp. Neurol. **236**(2), 383–388 (2012). https://doi.org/10.1016/j.expneurol.2012.04.024
109. T. Mera, J.L. Vitek, J.L. Alberts, J.P. Giuffrida, Kinematic optimization of deep brain stimulation across multiple motor symptoms in Parkinson's disease. J. Neurosci. Methods **198**(2), 280–286 (2011). https://doi.org/10.1016/j.jneumeth.2011.03.019
110. C.R. Butson, S.E. Cooper, J.M. Henderson, B. Wolgamuth, C.C. McIntyre, Probabilistic analysis of activation volumes generated during deep brain stimulation. NeuroImage **54**(3), 2096–2104 (2011). https://doi.org/10.1016/j.neuroimage.2010.10.059
111. A. Eusebio, H. Cagnan, P. Brown, Does suppression of oscillatory synchronisation mediate some of the therapeutic effects of DBS in patients with Parkinson's disease? Front. Integr. Neurosci. **6**, 47 (2012). https://doi.org/10.3389/fnint.2012.00047
112. L.A. Johnson, S.D. Nebeck, A. Muralidharan, M.D. Johnson, K.B. Baker, J.L. Vitek, Closed-loop deep brain stimulation effects on parkinsonian motor symptoms in a non-human primate – is beta enough? Brain Stimul. **9**(6), 892–896 (2016). https://doi.org/10.1016/j.brs.2016.06.051
113. G. Kleiner-Fisman, D.N. Fisman, E. Sime, J.A. Saint-Cyr, A.M. Lozano, A.E. Lang, Long-term follow up of bilateral deep brain stimulation of the subthalamic nucleus in patients with advanced Parkinson disease. J. Neurosurg. **99**(3), 489–495 (2003). https://doi.org/10.3171/jns.2003.99.3.0489
114. E. Ryapolova-Webb, P. Afshar, S. Stanslaski, T. Denison, C. De Hemptinne, K. Bankiewicz, P.A. Starr, Chronic cortical and electromyographic recordings from a fully implantable device: preclinical experience in a nonhuman primate. J. Neural Eng. **11**(1), 016009 (2014). https://doi.org/10.1088/1741-2560/11/1/016009
115. P. Wen, M. Li, H. Xiao, R. Ding, H. Chen, J. Chang, et al., Low-frequency stimulation of the pedunculopontine nucleus affects gait and the neurotransmitter level in the ventrolateral thalamic nucleus in 6-OHDA Parkinsonian rats. Neurosci. Lett. **600**, 62–68 (2015). https://doi.org/10.1016/j.neulet.2015.06.006

Cognitive and Physiological Intent for the Adaptation of Motor Prostheses

Raviraj Nataraj, Sean Sanford, Mingxiao Liu, Kevin Walsh, Samuel Wilder, Anthony Santo, and David Hollinger

Abstract Motor prostheses act to restore function to persons with movement disability through device actions trigged by command or intent of the user. There are various modes by which device actions may power or inform the person's movement. For any device mode, actions are based on what the person is intending to do. Regardless of the device, user, or method for restoring function, it is critical to optimize how the device and user respond to one another to maximize the capabilities of the person. In this chapter, we discuss approaches to develop assistive devices and rehabilitation paradigms that restore function to those with movement disabilities through greater cognitive and physiological integration. Mainly, these approaches fall along two lines: (1) adapting devices to facilitate greater agency, or sense of control, of the user and (2) training a user to produce consistent responses that the device can readily recognize and assist.

Keywords Movement · Feedback control · Rehabilitation · Assistive devices · Cognitive agency · Device adaptation · Neuroprostheses · Exoskeletons

1 Introduction

The ability to stand, walk, and grasp can be severely impaired for persons with spinal cord injury and limb amputation [1, 2]. Advanced user-device interfaces for prosthetic control can restore the capabilities to perform activities of daily living for persons with severe movement dysfunction. State-of-the-art motor prostheses

---

R. Nataraj (✉) · S. Sanford · M. Liu · K. Walsh · S. Wilder · A. Santo · D. Hollinger
Movement Control Rehabilitation Laboratory, Biomedical Engineering Department, Stevens Institute of Technology, Hoboken, NJ, USA
e-mail: rnataraj@stevens.edu; ssanford@stevens.edu; mliu26@stevens.edu; kwalsh4@stevens.edu; swilder@stevens.edu; asanto@stevens.edu; dhollin1@stevens.edu

R. Vinjamuri (ed.), *Advances in Motor Neuroprostheses*,
https://doi.org/10.1007/978-3-030-38740-2_8

utilize increasingly sophisticated ways to integrate the user for more natural control and more effective function. Novel neural interfaces can directly access sensory and motor physiology of the person to trigger the operation of a powered device [3]. Recordings from motor pathways, such as electromyography signals, and activation of sensory pathways are typically used to better connect the person to their limb movements. These movements may then be powered by muscle activation [4] or device motors [5]. The ultimate objective for these prosthetic interfaces is to restore movement abilities that achieve normative performance and feel as natural as healthy limbs.

Despite major technological advancements for better movement actuation and greater access to user physiology for recording and activation, advanced motor prostheses are still not the standard of care. Clinical retention of these devices is challenged, not only by their expense but the daily cost in effort to deploy and use the device relative to the benefit of actual improvement in function. As such, it is critical to develop effective methods that empower the person to use the device with greater performance outcomes. The sense of empowerment of the user involves a greater perception of control and better matching the intentions of the user with the actions of the device. Cognitively engaging the person in the operation of the device should promote user empowerment toward greater functional performance.

Our lab is investigating cognitively driven approaches to better integrate a person having movement disability to a powered assistive device for greater rehabilitation gains. These approaches largely fall along two pathways: (1) adapting device operation to induce greater user agency and (2) more effectively training a user toward greater movement performance with a device. For (1), augmenting user agency over a device should facilitate greater cognitive engagement to device actions and subsequently generate greater performance. Operational parameters of the device (e.g., feedback gains of the control system) are classically optimized toward objectives of maximizing movement tracking or minimizing effort [6]. However, if reliable methods to measure and modulate agency are determined, then device parameters may also be tuned to maximize user integration to the device on a cognitive level. For (2), training a user to better command a device should generate greater synergistic user-device action for smoother control and better combined performance. Command interfaces for devices can be continually modified toward new training inputs [7]. But having the user effectively learn specific command patterns may further expand their own independent function. The combined ability of the user and device may be enhanced if the device can more readily interpret well-targeted user actions. We intend to explore multiple sensory modalities across vision, sound, proprioception, and tactile stimulation to serve as effective training guides that better integrate the intentions and actions of the user with those of the device.

2 Research Approach 1: Identifying Operational Conditions of Motor Prosthetic Devices That Enhance User Sense of Agency

In this section, we describe how individuals with neurotraumas or neural pathologies, which have compromised control and capacity of muscular actions, may benefit from cognitive-based rehabilitation. Individuals with this type of movement disability must either engage in rigorous physical therapy or utilize an assistive device to regain motor function. Physical therapy involves the repetitive practice of functional tasks such as grasping objects to reformulate neural connections and rebuild strength [8]. Physical therapy gains may be accelerated with the aid of technological interfaces such as virtual reality (VR) [9], robotic interfaces [5], and instrumented wearables [10]. These interfaces can provide systematic feedback to a user and the clinical team about performance and generate environments that are cognitively engaging to the person receiving therapy. Assistive device technology could also be used to provide functional assistance that could not otherwise be regained such as neuroprostheses after severe spinal cord injury [11] or amputation [12]. In the remainder of this section, we make the case of how those with movement disabilities following neurotrauma may benefit from rehabilitative technologies that include greater cognitive focus. We describe approaches that quantitatively capture the sense of agency such that parameters of devices could be adapted accordingly.

2.1 Neuromuscular Disability Can Lead to a Sense of Disengagement From One's Own Body

Neuromuscular traumas and pathologies can severely restrict people's motor function and their subsequent ability to perform activities of daily living (ADL). Individuals with stroke, spinal cord injury (SCI), traumatic brain injury (TBI), and amputation can have compromised walking, standing, and grasping function. Each year, approximately 800,000 people suffer a stroke [13], 17,700 people have a SCI [14], 1.5 million sustain a TBI [15], and 1.6 million people undergo an amputation in the United States alone [2]. Currently, there are four major categories of treatment for neuromuscular dysfunction: (1) medication to reduce pain and improve neuroactivity, (2) surgery to repair neurological or orthopedic traumas, (3) physical training to rebuild muscle strength and retrain motor skills, and (4) developing assistive devices such as prosthetics and exoskeletons to help patients to perform functional tasks. With rehabilitation methods involving training and assistive devices, the person must strongly participate in the process for motor recovery. However, delivery of these methods mainly focuses on direct physical restoration of motor function, but not necessarily cognitive engagement of the person. Neuromuscular pathology not only affects muscular coordination and capacity but also cognition of the movement. Disturbance of memory and executive functioning are common neurocognitive symptoms for TBI patients [16]. Stroke patients were found to suffer from cognitive impairment and dysfunction memory [17]. Individuals with SCI can develop feelings of depression and social abandonment [18]. Thus, engaging the person on a cognitive level should accelerate motor recovery. This engagement could be through building greater empowerment during the rehabilitation process and greater sense of control over the operation of an assistive device.

2.2 Sense of Agency and Its Consideration for Motor Prostheses to Rehabilitate Function

The sense of agency (SoA) refers to the feeling or the perception of being the true author of one's own actions and related consequences [19]. The SoA has been a significant target metric in experiments that investigate the cognitive link between action and outcomes. Experiments include those for movement actions such as the performance of reaching and grasping tasks [20, 21]. Reaching and grasping are common functional movement tasks that allow engagement with a surrounding environment and are vital in performing ADL. Inducing greater agency to improve functional performance seems intuitive, but this concept has not been well demonstrated. Further studies should investigate methods to co-maximize the SoA and functional performance for neuromuscular patients as a basis to accelerate motor rehabilitation. Systematic methods that establish a clear link between SoA and performance could be leveraged toward neuromuscular rehabilitation involving training paradigms or assistive devices. The SoA indicates the cognitive perception of control over one's actions and the related consequences. As such, the SoA could provide the foundation for cognitive integration between a person with a movement disability and related functional performance during rehabilitation paradigms or in using assistive devices.

Previous studies have investigated how technological interfaces could affect the SoA. In a study conducted by Coyle et al. [22], the SoA was found to be stronger for a novel skin-input controlled keyboard. A study by Limerick et al. [23] found that the SoA was weaker for speech interfaces compared to movement interfaces, further supporting the potential for agency-based movement rehabilitation. Berberian et al. [24] showed that the increased level of autopiloting of a device reduces the SoA, suggesting how human-machine interfaces for rehabilitation should include more continuous input from the user. Often, assistive devices such as exoskeletons trigger automatic actions once a user command exceeds a specific threshold. An example is a powered exoskeleton for gait whereby the users tilt their hips to trigger the exoskeleton to take an automatic step [25]. The user involvement is disjointed as the user is left to ride the step to its completion until he/she is ready to take another step. While cognitive agency has not been extensively considered for rehabilitation, cognitive-based approaches to rehabilitation have been examined.

Dibilio et al. [26] applied methods to enhance attention as a bridge for better gait performance following Parkinsons' disease. Ritterband-Rosenbaum et al. [27] did demonstrate that computer-training can have a positive effect on the SoA in children with spastic cerebral palsy. Overall, cognitive metrics such as the SoA could be a fruitful pathway for device-centered rehabilitation assuming metrics are derived reliably for translation to systematic changes in device operation.

For motor prostheses, it is typically not only actions of the device that need be considered for restoring function but also how the device may be embodied by the user. Embodiment is facilitated through the perception that the entity being moved is a natural physical extension of one's own body [28]. Yet, there is a varying dependence between agency and embodiment. In her study, on the one hand, Caspar et al. [28] indicated that while interacting with an intermediary that is congruent to the action, our SoA does not become reduced. On the other hand, an intermediary that is incongruent to the action performed by the owner does not necessarily reduce agency. This suggests that for motor prostheses, it is truly the sense of control over the action (agency) rather than the sense of physical connection that drives cognitively integrated function. While prosthetics or exoskeletons that better mimic natural physical structure can facilitate a better sense of connection to the user, it is agency over the device that binds the user to function. Caspar's results further showed that the SoA is flexible with the use of intermediaries such as a robotic hand. This finding further suggests that methods to systematically adapt a motor prosthesis according to agency may be valuable and highly plausible.

2.3 Linking Agency to Greater Movement Performance

Several approaches to measure SoA have been investigated, but these approaches are typically classified as either implicit or explicit [19]. Explicit measures generally require the subject to provide a survey-like response as to their perception of control during an experimental paradigm. The subject self-assesses to what extent their actions are responsible for the resulting actions such as with visual avatars, presumed to act under user command [29]. However, intentional binding, an implicit measure, is emerging as a reliable standard for measuring agency [30]. With intentional binding, greater agency is implied through greater temporal coupling of a perceived voluntary action and the intended consequence. A typical experimental consequence is a sound event occurring at some time interval after an action (e.g., keypress, [31]). It has been well demonstrated that actions which are more intentional or voluntary lead to greater compression in the perception of the time interval between the action and sensory event [30, 32, 33]. Our lab has developed experimental paradigms that vary visualized conditions of operation for a virtual reach-to-grasp task to investigate the correlation between sense of agency and performance (Fig. 1). Agency is assessed according to the subject's perception of time interval between completion of their grasp action and receiving a sound event (beep) as a sensory consequence. The operational conditions include variation in

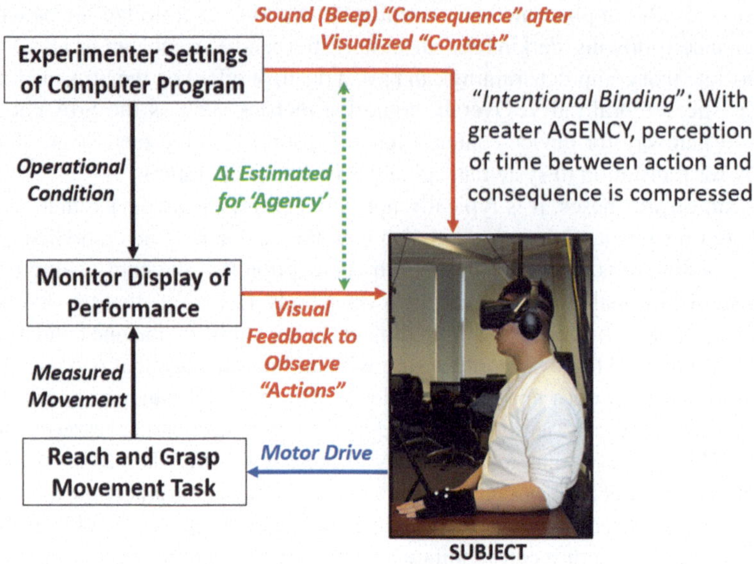

Fig. 1 Flow diagram of experiment having subject perform reach and grasp under varying visualized operation in virtual reality while assessing agency and performance

speed, inclusion of mild noise, and adding a level of automation to the subject's motor drive. These conditions serve to represent general operational levels by which to tune a device. These ongoing studies are establishing the significant and positive correlation between performance and agency for both reaching motion tasks [34] and grasp force-tracking tasks [35]. Our developing results strongly suggest that tuning operation of devices according to agency may be possible and can produce greater user-device integration for better cognition and performance.

2.4 Physiological Patterns as Implicit Measures for Agency

Explicit measurements of SoA are highly subjective, and while the SoA is a metric about perception, soliciting survey-like responses from a subject can be prone to various bias. It becomes more challenging to use such a measure to objectively tune an assistive device or rehabilitation paradigm toward greater performance. The tuning process effectively reduces to adapting operation of the interface toward a user's best guess of preference. This approach does not necessarily serve any further objective such as cognitive integration or better performance. While an implicit measure, intentional binding offers a more objective metric of SoA by having the subject focus on time estimation, a metric without qualitative inference. As with any cognitive-level metric, several trials for a given operational condition may be needed for reliable representation of a subject's SoA for that condition. However, a quantitative value can be assessed, and advanced processes such as mathematical optimization can be employed to identify control parameters automatically [36]. However, necessitating subjects to focus on the additional task of providing a quantitative metric in addition to performing the desired task may be fatiguing. In general, fatigue has cognitive repercussions [37] and may adversely affect the user-device "fitting" process. Finding ways by which SoA may be inferred more passively such that the subject simply engages in the process of using the device at a given operating condition is the most desirable. Other implicit measures for agency may need to be found that ascertain unconscious subject responses such as eye movements [38] or subtle gestures [39] while tuning devices.

Our lab is investigating physiological responses such as electromyography (EMG, Delsys, Trigno system) and electroencephalography (EEG, 32-channel *g.tec* system) as indicators for higher SoA toward greater movement performance. EMG and EEG are highly implicated as control signals for movement assistance devices [40–43]. Our expectation is that the standard method of intentional binding to measure the SoA can be used as "validation" to find reliable neural signatures through EEG-EMG coherence during motor function tasks (Fig. 2). Motion (*Optitrack*, nine *Prime 17 W* cameras) and force (*AMTI, Mini40*) capture was processed in real time to provide performance cues to the subject performing a grasp force-tracking task. Kühn et al. [44] showed that the lateral, caudal region within the supplementary motor complex strongly contributes to the experience of SoA when performing a voluntary action. It was also shown that specific bands (alpha, beta) may contain the primary neural oscillations associated with SoA [40]. Sato et al. demonstrated that high SoA may be experienced in myoelectric control of a robotic arm when modulating the visualized synchrony between one's own actions and those of the robotic arm [45]. As such, if physiological foundations for agency could be reliably identified, then automatic adaptation of the device for greater integration to the user could be developed for a variety of human-machine interfaces, including motor prostheses.

2.5 Utilizing Reward-Based Rehabilitation to Enhance Agency and Performance

Training approaches to rehabilitate individuals with neuromotor disability should consider incentivized ways in which to provide performance feedback. It has been shown that virtual reality rehabilitation can produce gains over conventional physical therapy [45]. Not only are the interfaces more visually stimulating, but training paradigms may be presented in gamified ways to further motivate engagement [46]. When the feedback is positive and interpreted as "reward," it can stimulate adaptations in neural processes toward greater retention of better performance and more independent function [47]. As such, reward may not only facilitate motor learning but it may also be a vehicle from which to cognitively integrate users and movement devices.

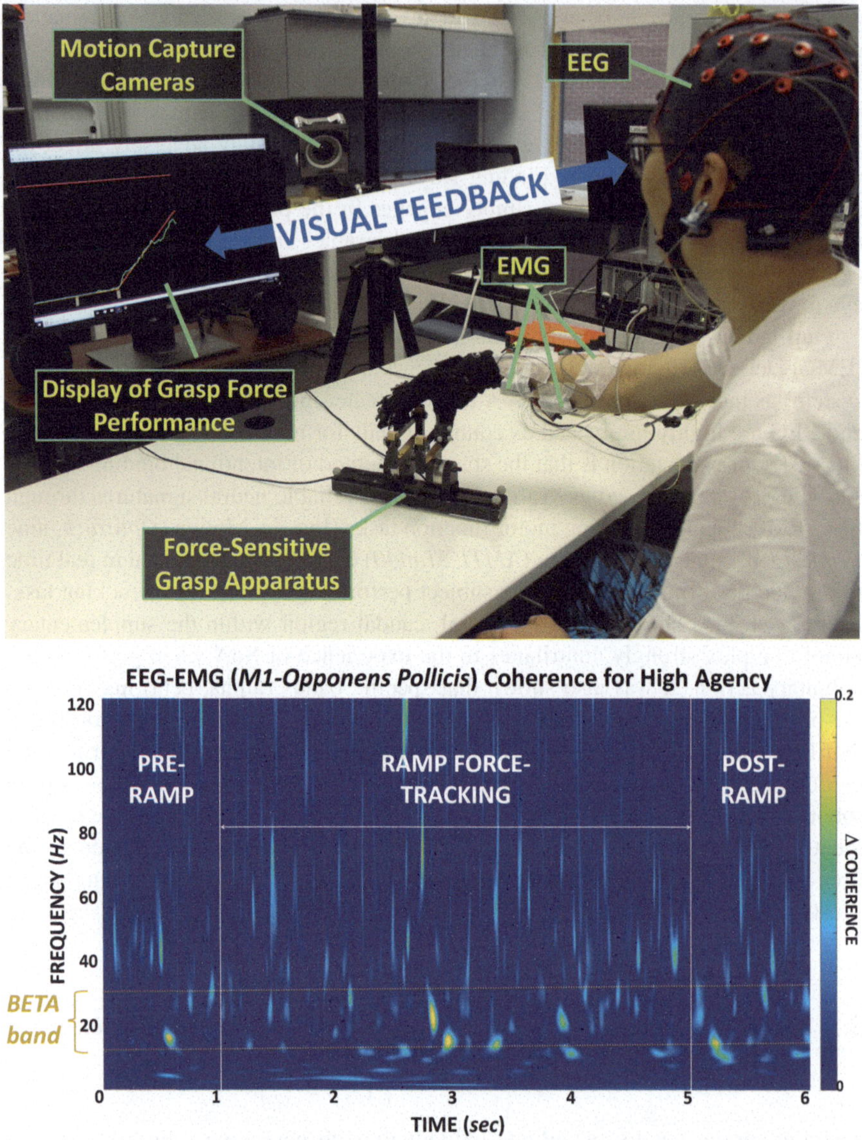

Fig. 2 Paradigm to observe physiological underpinnings of agency during movement task. Top—Subject grasping instrumented apparatus during task to force-track a target ramp. Bottom—Difference in coherence for select EEG-EMG pair between trials recorded as high agency versus low agency

The effect of reward to motivate movement performance has been demonstrated extensively. Mulliken et al. [48] designed a series of experiments using juice as a reward to motivate monkeys to rapidly learn optimal control of a screen cursor. Brain recordings indicated a higher potential to decode those neural signals for better neuroprosthetics. Another study investigated how reward with virtual reality rehabilitation could amplify movement magnitudes during reaching of the affected hand following hemiparesis [49]. As such, visual projections of performance that promote a sense of success appear to accelerate potential rehabilitation gains. Amputees learning to operate an external device can leverage a similar approach to operate the device skillfully for ADL [2].

Despite the apparent benefits in providing reward with performance feedback for rehabilitation, it is not clear what type or dosage of reward would optimally accelerate gains or best integrate user actions to an assistive device. Our lab has been investigating the use of simple reward and punishment feedback on performance and agency of a virtual reach-to-touch task (Fig. 3). Preliminary results have indicated that regardless of actual performance when receiving the feedback, higher dosages of reward compared to punishment produce higher agency and performance [50, 51]. Furthermore, a general positive reward may not have the same effect on each individual. A personality type where one is more driven by extrinsic cues [52] may have a greater sensitivity to either reward or punishment feedback. Also, providing visual cues with greater positive familiarity (e.g., imagery associated with hobbies) may also produce desirable neural adaptations [53] that could be leveraged for greater cognitive integration between user and rehabilitation device.

2.6 Sensory Feedback to Induce Greater Agency

In addition to perceptive feedback (e.g., reward), explicit sensory feedback may have an important role in inducing greater agency and performance of a person operating a movement device. Visual feedback has been widely demonstrated to be useful in informing individuals toward improved movement performance [54]. Visual feedback is also a dominating cue for SoA. Evans et al. [55] showed that visual feedback overrides SoA when no proprioceptive feedback is provided. Methods by which sensory feedback can be systematically delivered to achieve specific objectives such as higher agency and greater performance would be valuable for user-device integration of motor prostheses. Virtual reality is a powerful environment from which to provide highly customized visual feedback at a low cost for motor rehabilitation training [9, 46]. Audio, haptic, and tactile cues could further reinforce visual feedback toward performance and agency during user-device integration. When presented appropriately, these various feedback sources have already been demonstrated to positively affect cognitive behavior [55] and functional touch sensation [56] following neurotrauma.

In our lab, we are examining methods to leverage various sensory feedback to induce greater agency toward higher performance. We are developing an instru-

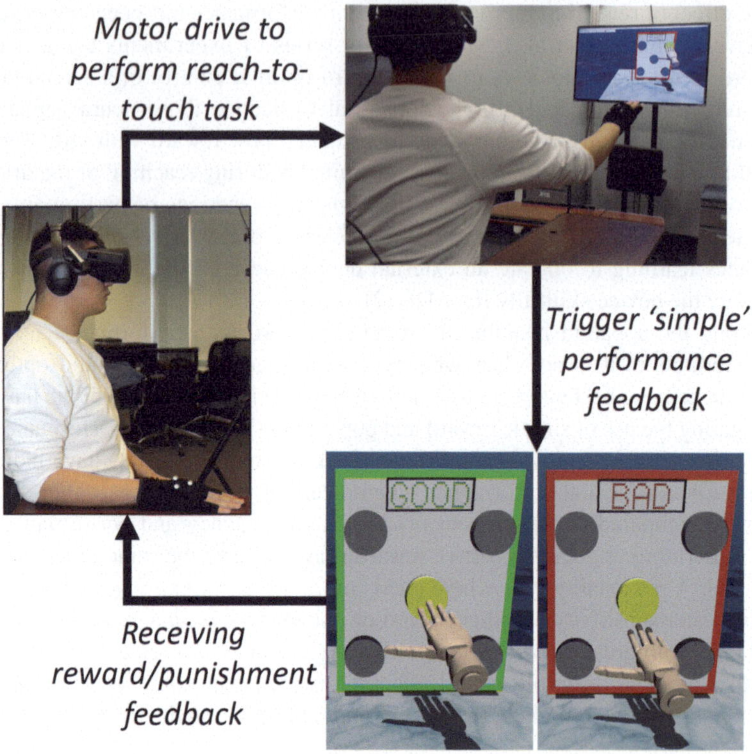

Fig. 3 Example of performing multi-DOF task in VR environment (*MuJoco HAPTIX*) with reward/punishment feedback on performance

mented glove, *The Cognition Glove,* which aims to induce greater SoA through the systematic provision of sensory feedback in response to the achievement of secure grasp [35]. Secure grasp is accurately predicted using machine learning (ML) techniques to process signals from onboard flex and force sensors (Fig. 4). Given our procedures for generating training data, a simple artificial neural network was sufficient to identify true positive cases of secure grasp at a 90% rate for a variety of grip postures (whole-hand, tripod, precision pinch). Sensory cues are provided at variable timings under various modes (visual, LED; auditory, beeper; tactile, vibrator) to progressively induce agency and more efficiently garner performance gains. Our lab will be performing clinical testing on individuals with traumatic brain injury using this platform to verify how progressive cognitive binding to performance may improve motor retention following neurotrauma. Using relatively low-cost, easy-to-use, wearable technology should provide a pathway for clinical translation for technologies that leverage user cognition to better movement performance. Such platforms may also be developed for training better control of a variety of movement devices.

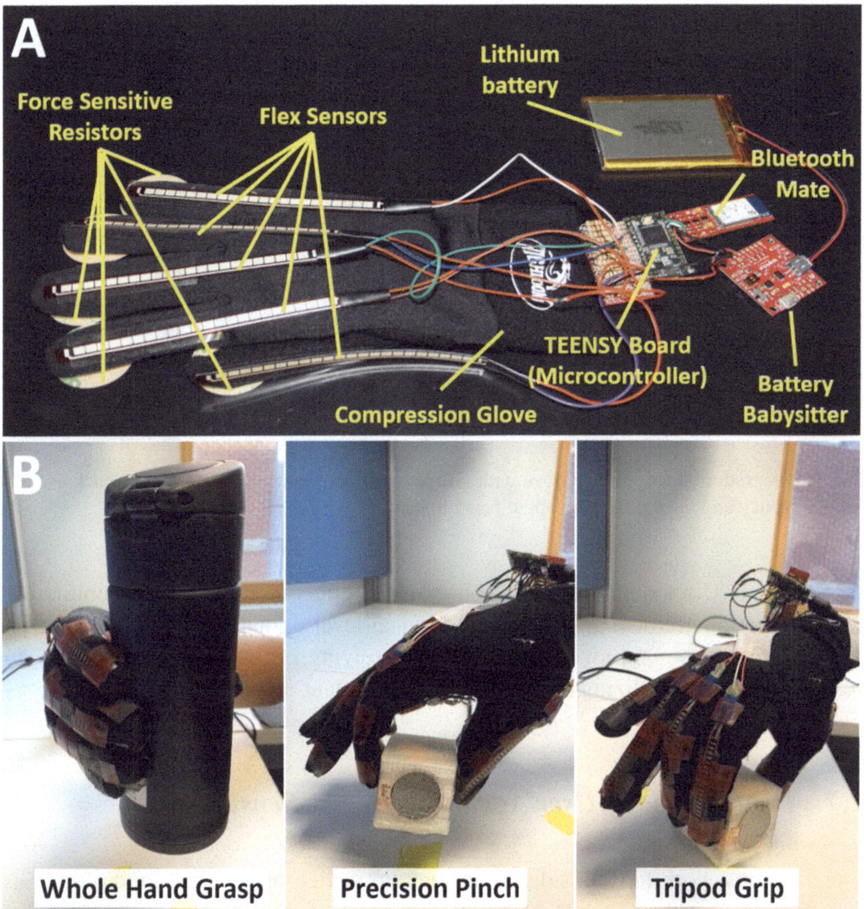

Fig. 4 Instrumented wearable glove ("Cognition Glove") informs the user about successful performance (secure grasp) with sensory feedback to induce greater agency. (**a**) Breakout of Cognition Glove, (**b**) Person using Cognition Glove for various grasp postures

Stand-alone wearable instrumentation providing milder cues would allow for cognitive-based training feedback throughout the day. However, optimal rehabilitation regimes would likely also include short periods of intensive training with "enhanced" sensory feedback. Combining instrumented wearables with virtual reality systems would offer a pathway to provide "enhanced" feedback while using technology that is highly affordable and portable for increased clinical translation. Our lab has developed VR applications (*Unity Technologies*) to integrate our *Cognition Glove* to commercially available VR technology such as the LEAP® for hand tracking and the Oculus® for immersion and visualization (Fig. 5). Our short-term objective is to investigate how enhanced sensory feedback from complementary VR could further augment short-term gain and retention in agency and

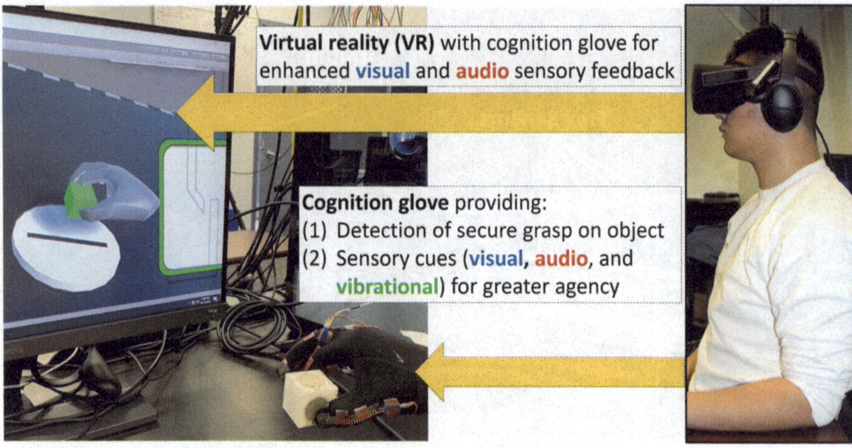

Fig. 5 Integration of Cognition Glove with virtual reality for enhanced sensory feedback used to induce agency and inform the user about performance

performance. Ultimately, we seek to observe this effect in individuals rehabilitating motor function following incomplete cervical-level spinal cord injury. Such an approach could be employed for individuals relying on physical therapy only to restore function or utilizing an assistive device such as a neuroprosthesis [12].

2.7 Dependence of User Agency to Device Sensitivity

Our previous work has considered changes in device operational modes such as fast or slow speed, inclusion of noise, and level of automation. However, for a given operational mode, the sensitivity at the command interface may also play an important role in optimizing performance and user agency over a device. For a motor prosthesis, command sensitivity in each of 6 degrees of freedom (6-DOF) may be adapted accordingly. It has been previously shown that increasing the gain on visualized speed will cause an increase in experienced agency [57]. Other studies have shown how command-interface sensitivity is crucial for driving control of devices for rehabilitation including hand exoskeletons [58] and wheelchairs [59].

A major objective for our lab is to develop methods that identify device sensitivity levels for user-device interfaces that optimize 6-DOF control through greater agency. Whether agency is identified through intentional binding or physiological metrics, we expect to demonstrate fundamental relationships between device sensitivity and agency that can be leveraged toward greater functional performance. At this stage, it is not clear whether these relationships are higher or lower order, linear or nonlinear, or uniquely dependent on a specific position or orientation DOF. It is also critical to determine for each application what window of test samples of a given sensitivity must be presented for the user to accommodate such that a reliable measure of agency is discerned.

Beyond procedural verification of how best to ascertain a reliable agency measurement, we are investigating how device sensitivity adaptation may be optimized for greater agency and performance across: (1) fixed versus flexible interfaces and (2) simple versus complex tasks (Fig. 6). In relation to motor prostheses, fixed interfaces would mimic myoelectric devices controlled using isometric contractions [60]. Flexible interfaces allow the user to physically move the point of interaction in commanding device operation. We hypothesize that the sense of agency may be heightened when the user has kinesthetic sensation [55, 61, 62] as part of the device control. We are developing tuning procedures for a commercially available 6-DOF mouse (*3DConnexion Spacemouse*) serving as a fixed interface. The user manipulates 6-DOF control within a computerized environment by applying loads upon the largely stationary mouse. For a flexible interface, a custom-built Gough Stewart platform [63] joystick was used. The user can manipulate each of three position and three orientation DOFs on the device whose corresponding sensor signals can be analogously translated to a computerized environment.

With both the flexible and fixed interface, we aim to test the efficacy of agency-based adaptation of command sensitivities with both simple and complex tasks. Full movement control would potentially involve manipulation of functional movement in all six DOFs. Control in more DOFs is interpreted as a complex user-device task. However, motor prostheses often restrict motions to a single plane with single-threshold command execution of the device to theoretically serve as a simple task. For example, powered exoskeletons for gait typically restricts leg motions to flexion-extension at the hips and knees following a single hip-tilt trigger. While exo-gait is a complex task with the user making 3-D balance adjustment with a crutch-cane, the trigger execution of a device-driven step is simple in theory. We aim to simulate *simple* (tracing of 2-D circle) and *complex* (6-DOF control of virtual hand) task functions for optimizing agency-based tuning of both fixed and flexible interfaces.

3 Research Approach 2: Utilizing Sensory Feedback to Train Consistent Movement Responses for Better Rehabilitation and Improved Use of Motor Prostheses

In this section, we discuss approaches and implications in using various modes of sensory feedback to better train a person with movement disability to use a myoelectric motor prosthesis. First, we make the case as to why training the user to produce consistent command responses is beneficial for device control. Next, we discuss various modes of sensory feedback that can be used for rehabilitation training to perform better movement patterns. We suggest how these approaches could be employed for better user integration to a myoelectric prosthesis. The effects on training and retention of various types of sensory feedback are considered. Then, we discuss how sensory feedback may be more effectively employed to train a user driving operation of a myoelectric device governed by machine learning decoding and control. Finally, we introduce the platform our lab is developing to utilize multisensory feedback to train and condition the user to produce responses that are more machine predictable. The goal of this platform is more efficient and effective user-device integration through training that includes VR simulation prior to deployment with real-world prosthetic systems.

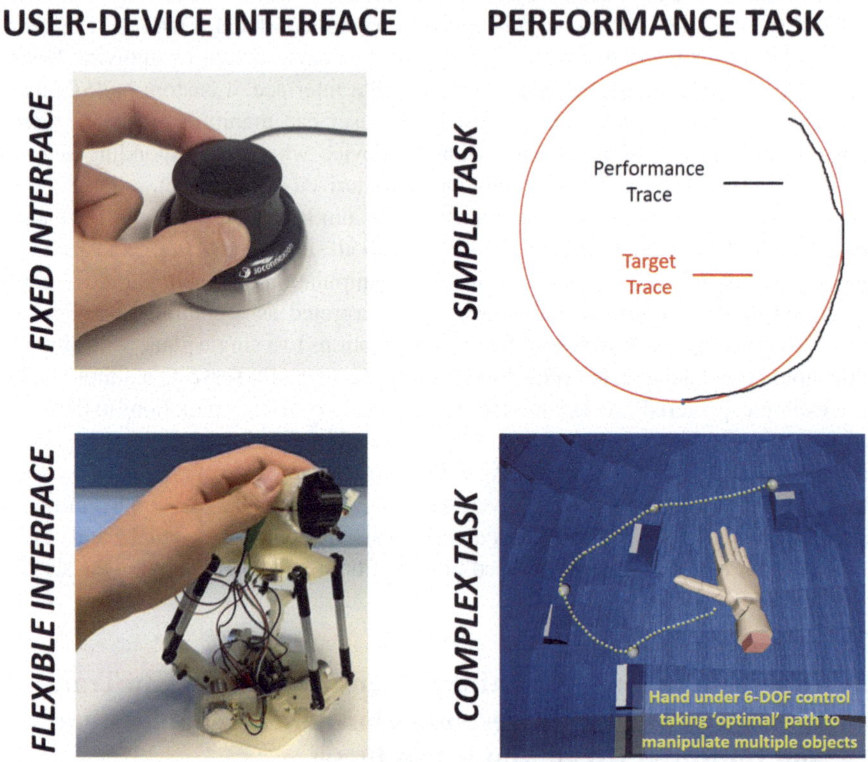

Fig. 6 Fixed and flexible interfaces serve as testing environments for varying device sensitivities and observing subsequent changes in agency and performance. These interfaces are used for performing either simple (2-D tracing) or complex (6-DOF) control tasks

3.1 Rationale to Training the User for Better Integration to a Myoelectric Motor Prosthesis

Following orthopedic trauma, rehabilitation involves effectively training a person to rebuild independent strength and coordination. In doing so, the objective is to produce responses that are accurate and consistent. Such training methods also prove beneficial in training the use of a motor prosthesis after neurotrauma, such as stroke, spinal cord injury, or amputation. In these cases, myoelectric control of an assistive device to restore function [64–66] may employ machine learning techniques that "adapt" device operation to user EMG command inputs [67, 68]. In theory, this approach is advantageous in that the onus is not on the user but rather the device to progressively alter behavior according to what the user is newly presenting as commands. Adaptation to the user is sensible given time-varying considerations of electrode placement, fatigue, or new muscle contraction tendencies with progressive use. However, the user inevitably must also endure this adaptation process. The goal in user-device integration is to achieve a reliable convergence in the behavior of each. With improved user ability to consistently generate target responses, the better the device will be able to reliably decode intent. In turn, the user can then develop agency over the device by better anticipating its actions. As such, exploring ways in which the user can learn to generate consistent command patterns would facilitate greater user-device integration.

Having the user generate targeted patterns would serve three sub-objectives. First, the user would be more actively engaged, at a cognitive level, in the user-device integration process by following guided cues as opposed to passively presenting various command patterns. Second, the adaptation process should be shorter if both the user and device have common targets upon which both can converge. The user and device would have a common target in functional directives (e.g., the movement directions) and respective commands the user is expected to give and the device is expected to receive. Third, the functional space is potentially larger as the command targets the user would be presented for training could be strategically more distinct and more easily discernible by a command pattern classifier. If left wholly to the unmotivated user, there could be tendencies to generate similar, overlapping command patterns. In which case, the machine learning classifier may be less able to find distinct muscle synergy for each unique command.

3.2 Sensory Feedback for Movement Training

While five senses are traditionally recognized, it is vision, sound, touch, proprioception, and vestibular feedback that are most implicated with completing ADL involving locomotion. Vision is dominant in real-time monitoring of one's movement relative to the surrounding environment. Sound provides cues about physical contact or verbal directives. Touch feedback informs one about the surface interaction of objects within the environment. While vision, sound, and tactile feedback are largely driven through external stimuli, proprioceptive (including kinesthesia) and vestibular feedback generate internal representations of body position and movement.

While those with neurotrauma may have reduced or eliminated modes of sensory perception, it is important to consider how sensation may be effectively leveraged toward better rehabilitation. The goal is to train the user, independently or with a motor prosthesis, to generate optimal movements for specific functions. Optimality criteria may include less effort, faster task completion, smoother movements, and reducing the onset of injury progression. In isolation, visual and haptic feedback are proven to generate better performance than audio cues in training movement [69, 70]. However, audio feedback can be used to effectively pace a cyclic activity [71] and provide sonification error to a movement variable [72] with minimal distraction. Visual feedback is the most effective sensory mode for guiding spatial positioning during movement tasks [73]. It typically involves the presentation of subject performance against a target to minimize error [73]. Tactile feedback involves activation of skin receptors, sensitive to pressure or vibration. Tactile feedback can be effective reinforcement cues to proprioceptive and vestibular tasks of controlling positions [74–76], regulating force [77–79], and maintaining multi-joint posture and balance [80–83].

3.3 Strategic Features in Sensory Feedback Training of Movement

While each type of sensory feedback mode may have unique positive features for movement training, advanced rehabilitation paradigms need to consider how each mode is presented for optimal effects. For example, visual feedback about performance can be provided in abstract forms whereby feedback "appears" as an extrinsic variable of performance rather than an intrinsic representation of one's own body movements. Examples of abstract feedback include simple plots, gauges, bars, or numbers [84–89]. Alternatively, visual feedback may also include more embodied representations as with virtual reality, 3-D animations, or virtual mirrors of one's own body [90–92]. Beyond the level of embodiment, other strategic considerations in how sensory feedback is provided for movement performance include the following: whether sensory feedback is beneficial more in real time or in retention, the complexity of the feedback, and how could various sensory modalities be effectively combined to train better user-device integration.

3.3.1 Sensory Feedback Training for Real-Time Performance Versus Retention

As a functional specification for a rehabilitation paradigm, it must be determined whether the goal of training for better movement is under real-time conditions or during retention. Typically, better independent function during retention, i.e., when feedback is removed, is the training objective. However, when operating a device, it may be plausible to continually include sensory feedback during ADL assuming the feedback is not considered noxious or distracting. *Concurrent* feedback is defined as real-time feedback of performance to immediately reduce error to the target. Concurrent feedback is highly beneficial in the early stages of learning to introduce the task objective and results in the highest level of performance [86]. However, concurrent feedback can be especially counterproductive for retention if it continuously over-cues the person. The guidance hypothesis [93, 94] suggests that constantly providing too much feedback is detrimental for retention because learning effects are more externally driven. As such, the person is not as encouraged to independently develop intrinsic mechanisms, such as muscle memory and proprioception. Furthermore, the performance during feedback training cannot be used to infer the effects on long-term learning [95].

To facilitate greater retention, sensory feedback training paradigms should reduce the dependency on feedback for movement execution and put more independent onus on the user. Previous approaches to this end include feedback that is faded [85, 96], self-driven [97–99], more implicit through softer bandwidth targets [100–102], and using terminal feedback [70, 103, 104]. *Terminal* feedback is when knowledge of results is provided immediately after performance task execution. The subject observes their performance offline in preparation of making improvements during subsequent trials. Intermittently providing both concurrent and terminal feedback in the same training blocks may offer both online performance and retention benefits [86]. Even if partner technology with the motor prosthesis would allow for some modes of continuous cueing, training approaches centered on promoting greater independent function is likely more desirable.

3.3.2 Feedback Complexity in Regulating Movement Performance

Simple tasks may be discriminated from complex ones according to the number of movement degrees of freedom involved. As such, complex feedback offering multiple concurrent targets can have positive effects with more complex tasks, especially when the subject is naïve to the task [105–108]. For tasks that involve multiple joints but virtually a single target objective, e.g., movement primarily in a single dimension, it is not clear how feedback complexity can optimally contribute to movement training. In our lab, we investigated how visual feedback complexity may elicit improved performance of the two-legged squat. Specifically, we evaluated how changes in complexity and body discernibility had an impact on the consistency of the subject's thigh position. The two-legged squat was a desirable platform

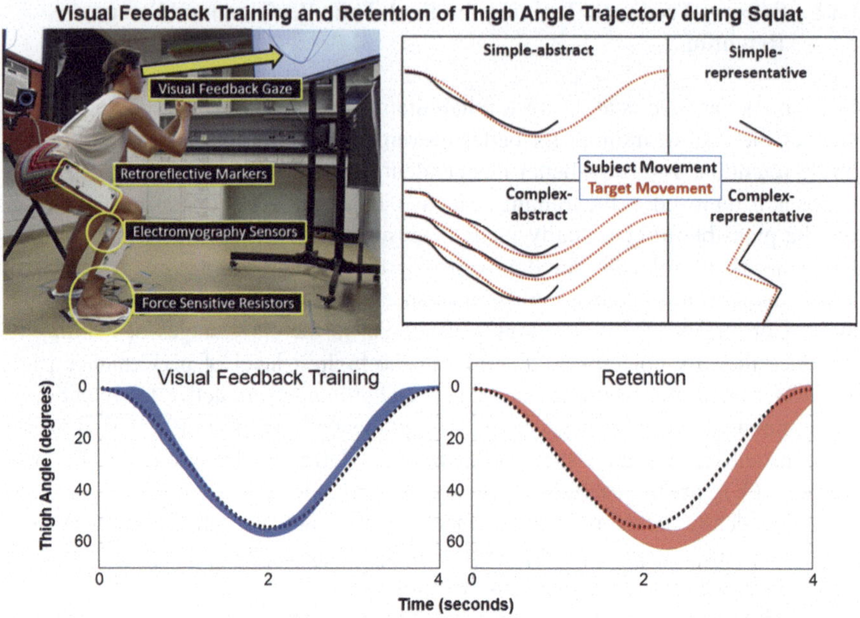

Fig. 7 Experimental setup provides visual feedback (VF) to the subject performing the two-legged squat exercise. Top—Subject observes one of four VF cases with varying complexity and body representation. Bottom—Sample performance traces during VF training versus retention trials. Performance trace is given as the subject's mean thigh angle trajectory +/− one standard deviation

movement for this investigation since it is a multi-joint movement with essentially a single variable of modulation, i.e., squat depth. Furthermore, the squat has similar features to the sit-to-stand movement that can be powered by motor prostheses following spinal cord injury [109].

Real-time visual feedback (VF) was presented as either simple (thigh position only) or complex (shank, thigh, and torso positions) type and cross-presented again in either an abstract or representative (body-discernible) mode (Fig. 7). Subjects completed a concurrent visual feedback training block immediately followed by a retention test. Our results indicated that complex VF was the most effective case for real-time performance when presented with body-discernible visualization. Subjects had similar performance across both simple VF cases, suggesting body discernibility effects only take hold with multi-segment representations. Complex VF also showed better performance in retention than both simple VF cases. We postulate that complex-representative VF allowed subjects to reduce their focus explicitly on the thigh angle, thereby promoting greater development of intrinsic mechanisms.

3.3.3 Sensory Feedback Integration for Myoelectric Prosthetic Control

Following neurotrauma or pathology, both motor and sensory capabilities of the person may be severely impaired. While visual and auditory capabilities are usually maintained, losing the sense of "touch" and "feel of body" to carry out a task makes it difficult to skillfully modulate body interactions with the environment [110]. Some neuro-machine interfaces include the restoration of touch and kinesthetic sensation [111, 112]. The restoration of haptic feedback in myoelectric devices can aid in the accurate assessment of object shapes [113, 114] and locations in space [110]. Whether sensation is restored or relying only on residual sensation, utilizing haptic feedback along with visual or audio cues for movement training may generate the greater performance gains in rehabilitating functional movements. Sensory feedback integration establishes greater body ownership over a prosthetic limb [115–118], which facilitates greater user-device integration. With embodiment to a prosthetic device, the user can better perceive device actions as those of an intact limb. Perceptual physical ownership of the device can increase the sense of agency [19, 119, 120].

Thus, multisensory feedback during movement training would have the potential to produce profound physical and cognitive binding of the person to specific objectives. Again, these objectives may be either greater independent function or greater capabilities with an assistive device. In either case, multisensory training of the user to produce consistent and distinct patterns in muscle activity would be invaluable. In addition to visual and audio feedback, incorporating haptic feedback for movement training can increase the control space of myoelectric prosthetic performance [121–125] and better "realize the effect" of actions [126]. Haptic feedback can be effectively employed for motor learning and increasing user engagement to a device by modulating the user's voluntary recruitment of muscles while either resistance or assistance is felt during the movement [126–128].

3.4 Machine Learning for Movement Control and Intent Detection

Due to the variability in myoelectric signaling and recording, it is vital to develop robust methods for training a machine learning classifier for reliable movement control and detection of user intent. Classifier systems typically acquire EMG data from the user and transfer that data into movement actuation of an assistive device or feedback to the user as part of training independent function. These classifiers must decipher system and command information from "patterns" in EMG signals from actively contracting muscles [129–133]. System and command information include the identification of ideal movement trajectories and intended directions of movement [4, 134–137]. The control system with embedded classification can

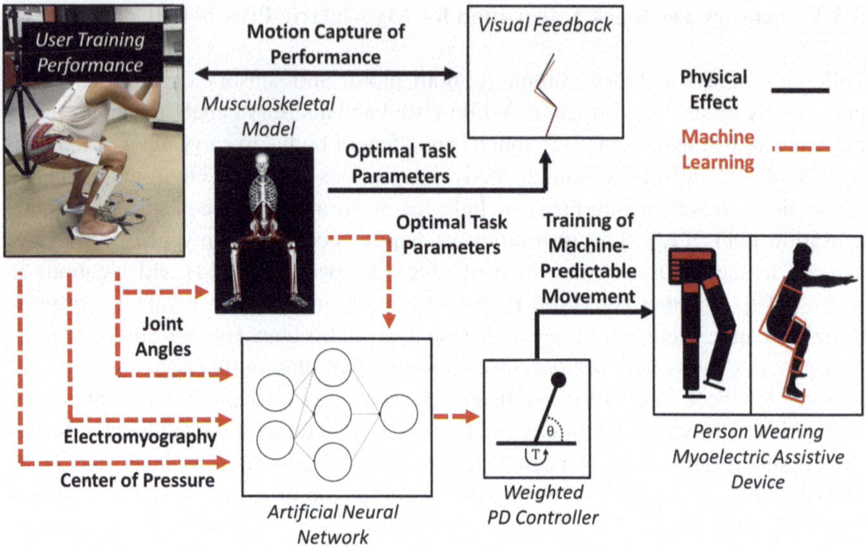

Fig. 8 Proposed "model" control system trains a user with visual feedback to collect data and adapt controller parameters of an assistive device

translate this EMG information to actuation signals that drive feedback displays or motors in the prosthetic device.

Our lab is developing systems in simulation to verify how well movement and EMG patterns collected during our squat VF experiments can be used to modulate squat performance of a biomechanical model that includes open-loop and closed-loop systems of control (Fig. 8). For each VF case, the open-loop component is based on actual subject data of motion (joint angles) and forces (ground reaction forces and center of pressure locations), from which joint moments are estimated. The closed-loop component includes additional joint torque drive from a neuro-PD controller [138–140]. Blana et al. [65] similarly used an artificial neural network inputting combined EMG and kinematic signals to control a prosthesis in a virtual reality environment during an upper limb reaching task.

The artificial neural network (ANN) receives input about the user motion, including EMG, and compares it against optimal (minimal effort) movement trajectories. These optimal trajectories are computed offline from the optimization of a musculoskeletal model for completing the squat from various user states. The ANN outputs joint errors that then drive a classical proportional-derivative (PD) control system to generate command torques onto the biomechanical model of the person and device. Parameters for the ANN (weights, biases) and the PD controller (proportional, derivative) may be optimized according to multiple dynamic simulations in which the squat is performed to minimize deviations from desired trajectories. While this model formulation does not consider how the open-loop component of the user may further change in the presence of closed-loop control, it offers the first approximation of possible performance under various VF training cases for user-device integration.

For the detection of commanded intention, many different machine learning classifiers have previously been applied to EMG signals to generate control inputs, including fuzzy logic [141], support vector machines [142, 143], linear and quadratic discriminant analysis [143], and artificial neural networks [65, 143]. EMG signals are recorded, and an ML classifier is trained on these signals to produce usable command inputs for the prosthetic or other device. Westwick et al. [144] generally observed that accuracy for EMG classification of movement increased when trained on additional recordings from intracortical electrodes. However, it was also observed that accuracy could increase further if the recording inputs were limited to "an appropriate selection of inputs." Westwick states that with the inclusion of too many neural inputs, cross-correlations between neurons produce a numerically ill-conditioned estimation and poor generalization of classification when presented with new data sets. Irrelevant or redundant additional inputs to a classifier can increase the training time and embed the classifier with a greater ability to reject noise rather than identify target signals. Furthermore, robust control of simple systems may be possible with only one input channel [145]. However, increasing the number of inputs generally allows control of more degrees of freedom and with greater accuracy if the classifier is robust enough to identify the underlying common-mode signal or pattern [144, 146, 147]. For example, ensemble ML methods perform very well with noisy data, such as neural signals. Ensemble methods cogenerate many learners, some of which may be weak but still contribute to the recognition of an evident pattern. Those learners will subsequently have greater influence on the ensemble's output. By recognizing several additional underlying patterns, greater discrimination between noise and true signals for command intention can be made across the ML ensemble [148]. We propose that useful patterns for a larger control space may emerge more reliably with the use of multisensory feedback to train the user.

We are conducting a study to observe classification accuracy for various muscle sets that have implications to respective neurological deficits [149]. Depending on the nature and extent of the deficit, e.g., level of SCI lesion, the available muscles for myoelectric control will vary. In our study, we expectedly observe the larger muscle sets producing better classification accuracy of the true-positive intention of directional control (Fig. 9). We are currently examining how these classifiers necessarily differ in performance when used for a functional task, such as control of a virtual object. We seek to determine if there is strict correlation between classification accuracy and performance and which minimal-input muscle sets can be reliably used moving forward when we seek to modulate those inputs with sensory feedback training.

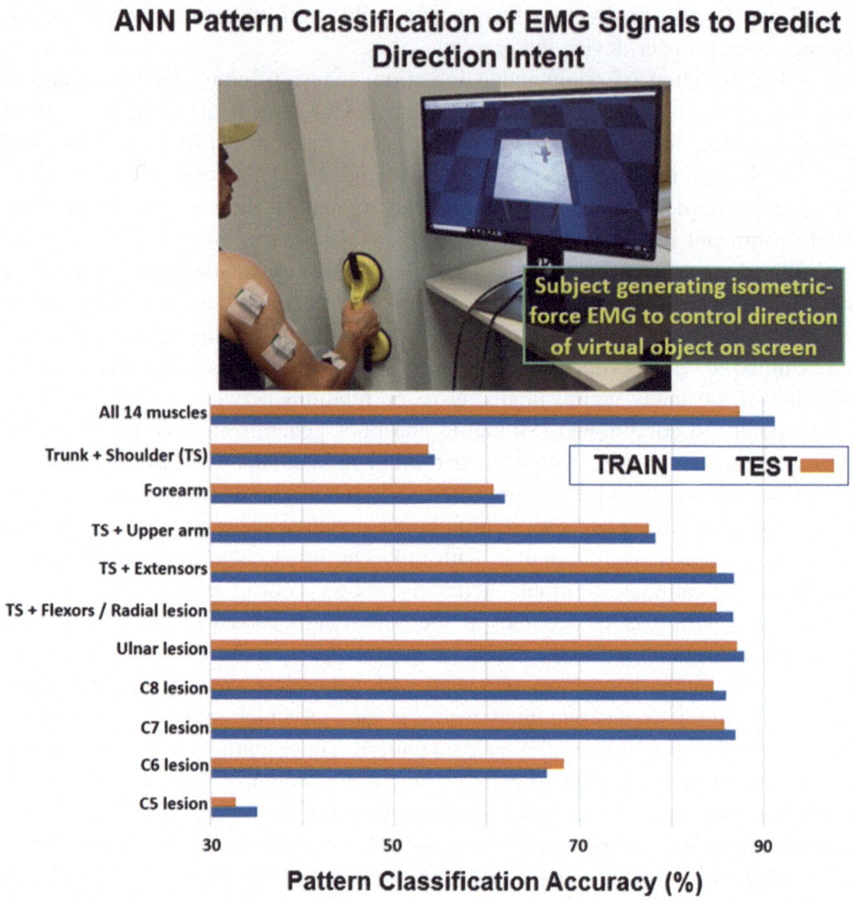

Fig. 9 Sensory feedback training paradigm for myoelectric device control

3.5 *Multisensory Platform to Train Users for Cognitive Integration to Motor Prostheses Under Myoelectric Control*

Our lab is currently developing a multisensory feedback platform to train users to produce electromyography patterns for better myoelectric device performance (Fig. 10). The platform will include the integration of visual, audio, and haptic (tactile and proprioceptive) feedback to guide the user to produce target muscle activation patterns. These patterns will initially be based on user-specific tendencies but then be parsed toward particular EMG targets. These targets will be specified to allow for greater classification accuracy in decoding intent across multiple dimension of control and various input muscle sets. The objective is to provide combinations of visual, audio, and haptic feedback during training to promote

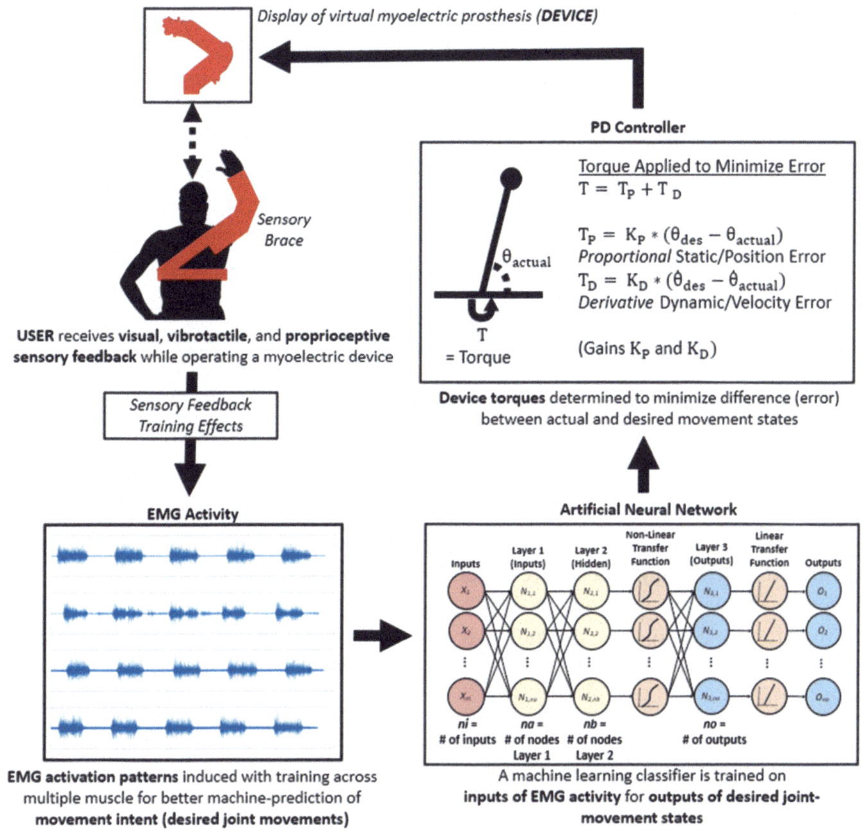

Fig. 10 Sensory feedback training platform for myoelectric device control

high online performance and development of intrinsic mechanisms. Visual and audio feedback will be provided primarily through a virtual reality interface also displaying the virtual prosthesis being controlled. Haptic feedback will be provided through vibration actuators and a worn brace that provides resistance to movement.

The upper-arm brace will provide either isometric or isodynamic (constant velocity) resistance and be constructed to include modules to integrate EMG sensors and vibrotactile stimulators. As the subject receives sensory feedback to drive the virtual prosthesis being observed, an ML technique, such as an artificial neural network, will take EMG inputs to continually generate the desired joint trajectories for the device. A proportional-derivative controller will continually compare these desired (command intended) trajectories to those being actually assumed by the virtual prosthesis to generate joint error signals. These joint error signals will be multiplied by the respective proportional and derivative feedback gains to produce the control torques to be applied to the virtual myoelectric prosthesis. The major engineering research objective with this platform is determining how multiple modalities of sensory feedback should be synergistically provided to best train the user. Considerations include feedback form (complexity, representation) and training instruction as to which feedback modes should the user explicitly pay attention versus be subconsciously guided. Ultimately, this platform will be used to determine optimal modes by which to present multisensory feedback and vary operational parameters for machine learning control and classification of intent. The platform is designed to be flexible to consider several clinical populations with varying levels of neuromotor deficit. While initially a simulation platform, it is expected to generate customized user-device parameters to be automatically ported to a real-world device.

4 Conclusions

In this chapter, we provide rationale and proposed approaches for the cognitive integration of users and motor prosthetic devices. We outline work from our lab as example methods by which to accomplish user-device integration in these ways. Mainly, the two major lines of research being promoted include the following: (1) adapting the device to augment user agency and (2) training the user to generate optimal device commands. As such, we propose methods that should facilitate faster and more effective convergence of operation for both the user and device from the perspective of each.

References

1. N. Suarez, R. Levi, J. Bullington, Regaining health and wellbeing after traumatic spinal cord injury. J. Rehabil. Med. **45**(10), 1023–1027 (2013)
2. P.F. Pasquina et al., Special considerations for multiple limb amputation. Curr. Phys. Med. Rehabil. Rep. **2**(4), 273–289 (2014)
3. D.P. Murphy et al., Electroencephalogram-based brain–computer interface and lower-limb prosthesis control: a case study. Front. Neurol. **8**, 696 (2017)
4. P.K. Artemiadis, K.J. Kyriakopoulos, EMG-based position and force control of a robot arm: application to teleoperation and orthosis. In *2007 IEEE/ASME International Conference on Advanced Intelligent Mechatronics*, Zurich, Switzerland (2007), pp. 1–6
5. J. Iqbal, H. Khan, N.G. Tsagarakis, D.G. Caldwell, A novel exoskeleton robotic system for hand rehabilitation – conceptualization to prototyping. Biocybern. Biomed. Eng. **34**(2), 79–89 (2014)
6. R. Nataraj, A.J. van den Bogert, Simulation analysis of linear quadratic regulator control of sagittal-plane human walking—implications for exoskeletons. J. Biomech. Eng. **139**(10), 101009 (2017)
7. B. Whitsell, P. Artemiadis, Physical Human–Robot Interaction (pHRI) in 6 DOF with asymmetric cooperation. IEEE Access **5**, 10834–10845 (2017)
8. S.M. Hunter et al., Functional strength training and movement performance therapy for upper limb recovery early poststroke—efficacy, neural correlates, predictive markers, and cost-effectiveness: FAST-INdiCATE trial. Front. Neurol. **8**, 733 (2018)
9. S.V. Adamovich et al., A virtual reality—based exercise system for hand rehabilitation post-stroke. Presence Teleoperators Virtual Environ. **14**(2), 161–174 (2005)

10. L. Connelly, Y. Jia, M.L. Toro, M.E. Stoykov, R.V. Kenyon, D.G. Kamper, A pneumatic glove and immersive virtual reality environment for hand rehabilitative training after stroke. IEEE Trans. Neural Syst. Rehabil. Eng. **18**(5), 551–559 (2010)
11. J.L. Collinger, S. Foldes, T.M. Bruns, B. Wodlinger, R. Gaunt, D.J. Weber, Neuroprosthetic technology for individuals with spinal cord injury. J. Spinal Cord Med. **36**(4), 258–272 (2013)
12. B.C. Eapen, D.P. Murphy, D.X. Cifu, Neuroprosthetics in amputee and brain injury rehabilitation. Exp. Neurol. **287**(Pt 4), 479–485 (2017)
13. Heart disease and stroke statistics—2017 update: a report from the American Heart Association, p. 458
14. Facts and Figures at a Glance. Birmingham, *National Spinal Cord Injury Statistical Center*, 2018 [Online]. Available from https://www.nscisc.uab.edu/Public/Facts%20and%20Figures%20-%202018.pdf. Accessed 28 Aug 2019
15. D.J. Thurman, C. Alverson, K.A. Dunn, J. Guerrero, J.E. Sniezek, Traumatic brain injury in the United States: a public health perspective. J. Head Trauma Rehabil. **14**(6), 602–615 (1999)
16. D.B. Arciniegas, K. Held, P. Wagner, Cognitive impairment following traumatic brain injury. Curr. Treat. Options Neurol. **4**(1), 43–57 (2002)
17. N.K. Al-Qazzaz, S.H. Ali, S.A. Ahmad, S. Islam, K. Mohamad, Cognitive impairment and memory dysfunction after a stroke diagnosis: a post-stroke memory assessment. Neuropsychiatr. Dis. Treat. **10**, 1677–1691 (2014)
18. Z. Khazaeipour, S.-M. Taheri-Otaghsara, M. Naghdi, Depression following spinal cord injury: its relationship to demographic and socioeconomic indicators. Top. Spinal Cord Inj. Rehabil. **21**(2), 149–155 (2015)
19. J.W. Moore, What is the sense of agency and why does it matter? Front. Psychol. **7**, 1272 (2016)
20. A. Sato, A. Yasuda, Illusion of sense of self-agency: discrepancy between the predicted and actual sensory consequences of actions modulates the sense of self-agency, but not the sense of self-ownership. Cognition **94**(3), 241–255 (2005)
21. S. Hamasaki, et al., Evaluating effect of sense of ownership and sense of agency on body representation change of human upper limb. In *2015 International Symposium on Micro-NanoMechatronics and Human Science (MHS)*, Nagoya, Japan (2015), pp. 1–5
22. D. Coyle, J. Moore, P.O. Kristensson, P. Fletcher, A. Blackwell, I did that!: measuring users' experience of agency in their own actions. In *Proceedings of the 2012 ACM Annual Conference on Human Factors in Computing Systems – CHI '12*, Austin, Texas, USA (2012), p. 2025
23. H. Limerick, J.W. Moore, D. Coyle, Empirical evidence for a diminished sense of agency in speech interfaces. In *Proceedings of the 33rd Annual ACM Conference on Human Factors in Computing Systems – CHI '15*, Seoul, Republic of Korea (2015), pp. 3967–3970
24. B. Berberian, J.-C. Sarrazin, P.L. Blaye, P. Haggard, Automation technology and sense of control: a window on human agency. PLoS One **7**(3), e34075 (2012)
25. R.J. Farris, Design of a powered lower-limb exoskeleton and control for gait assistance in paraplegics, p. 114
26. V. Dibilio et al., Computer-assisted cognitive rehabilitation on freezing of gait in Parkinson's disease: a pilot study. Neurosci. Lett. **654**, 38–41 (2017)
27. A. Ritterband-Rosenbaum, M.S. Christensen, J.B. Nielsen, Twenty weeks of computer-training improves sense of agency in children with spastic cerebral palsy. Res. Dev. Disabil. **33**(4), 1227–1234 (2012)
28. E.A. Caspar, A. Cleeremans, P. Haggard, The relationship between human agency and embodiment. Conscious. Cogn. **33**, 226–236 (2015)
29. S. Subramanian, L.A. Knaut, C. Beaudoin, B.J. McFadyen, A.G. Feldman, M.F. Levin, Virtual reality environments for post-stroke arm rehabilitation. J. NeuroEng. Rehabil. **4**(1), 20 (2007)
30. P. Haggard, S. Clark, J. Kalogeras, Voluntary action and conscious awareness. Nat. Neurosci. **5**(4), 382–385 (2002)

31. S.S. Obhi, P. Hall, Sense of agency and intentional binding in joint action. Exp. Brain Res. **211**(3–4), 655–662 (2011)
32. H. Limerick, D. Coyle, J.W. Moore, The experience of agency in human-computer interactions: a review. Front. Hum. Neurosci. **8**, 643 (2014)
33. N. Braun, J.D. Thorne, H. Hildebrandt, S. Debener, Interplay of agency and ownership: the intentional binding and rubber hand illusion paradigm combined. PLoS One **9**(11), e111967 (2014)
34. R. Nataraj, A. Shah, S. Sanford, Role of cognitive agency in reach-to-grasp movement performance, presented at the American Society of Biomechanics (August 2018), Rochester, MN (2018)
35. R. Nataraj, S. Sanford, M. Liu, S. Wilder, C. Rybarski, 16th International symposium on computer methods in biomechanics and biomedical engineering and the 4th conference on imaging and visualization abstract book, *E – 07 Biomech. Mov. Rehabil. Bioeng. II*, p. 251
36. I. Vierhaus, A. Fügenschuh, R.L. Gottwald, S.N. Groesser, Modern nonlinear optimization techniques for an optimal control of system dynamics models (2014)
37. K. Kahol et al., Effect of fatigue on psychomotor and cognitive skills. Am. J. Surg. **195**(2), 195–204 (2008)
38. Y.-F. Tsai, E. Viirre, C. Strychacz, B. Chase, T.-P. Jung, Task performance and eye activity: predicting behavior relating to cognitive workload. Aviat. Space Environ. Med. **78**(5 Suppl), B176–B185 (2007)
39. S. Helme, D. Clarke, Identifying cognitive engagement in the mathematics classroom. Math. Educ. Res. J. **13**(2), 133–153 (2001)
40. S.Y. Kang et al., Brain networks responsible for sense of agency: an EEG study. PLoS One **10**(8), e0135261 (2015)
41. T. Li, T. Xue, B. Wang, J. Zhang, Decoding voluntary movement of single hand based on analysis of brain connectivity by using EEG signals. Front. Hum. Neurosci. **12**, 381 (2018)
42. C. Antfolk, Using EMG for real time prediction of joint angles to control a prosthetic hand equipped with a sensory feedback system. J. Med. Biol. Eng. **30**(6), 399 (2010)
43. R.M. Singh, S. Chatterji, Trends and challenges in EMG based control scheme of exoskeleton robots – a review. Int. J. Sci. Eng. Res. **3**(8), 8 (2012)
44. S. Kühn, M. Brass, P. Haggard, Feeling in control: neural correlates of experience of agency. Cortex **49**(7), 1935–1942 (2013)
45. Y. Sato, T. Kawase, K. Takano, C. Spence, K. Kansaku, Body ownership and agency altered by an electromyographically controlled robotic arm. R. Soc. Open Sci. **5**(5), 172170 (2018)
46. F. Argelaguet, L. Hoyet, M. Trico, A. Lecuyer, The role of interaction in virtual embodiment: effects of the virtual hand representation. In *2016 IEEE Virtual Reality (VR)*, Greenville, SC, USA (2016), pp. 3–10
47. Y. Song, A.L. Smiley-Oyen, Probability differently modulating the effects of reward and punishment on visuomotor adaptation. Exp. Brain Res. **235**(12), 3605–3618 (2017)
48. G.H. Mulliken, S. Musallam, R.A. Andersen, Decoding trajectories from posterior parietal cortex ensembles. J. Neurosci. **28**(48), 12913–12926 (2008)
49. B.R. Ballester et al., The visual amplification of goal-oriented movements counteracts acquired non-use in hemiparetic stroke patients. J. NeuroEng. Rehabil. **12**(1), 50 (2015)
50. D. Hollinger, A. Shah, R. Nataraj, Accelerating neuromotor rehabilitation with reward feedback, presented at the Biomedical Engineering Society Annual Meeting (October 2018), Atlanta, GA (2018)
51. D. Hollinger, R. Nataraj, Acclerating neuromotor learning with reward feedback, Biomedical Engineering, Stevens Institute of Technology (2018)
52. P.S. Richardson, A.S. Dick, A.K. Jain, Extrinsic and intrinsic cue effects on perceptions of store brand quality. J. Mark. **58**(4), 28–36 (1994)
53. M.A. Webster, D.I.A. MacLeod, Visual adaptation and face perception. Philos. Trans. R. Soc. B: Biol. Sci. **366**(1571), 1702–1725 (2011)
54. J.J. Triano, J. Scaringe, J. Bougie, C. Rogers, Effects of visual feedback on manipulation performance and patient ratings. J. Manip. Physiol. Ther. **29**(5), 378–385 (2006)

55. N. Evans, S. Gale, A. Schurger, O. Blanke, Visual feedback dominates the sense of agency for brain-machine actions. PLoS One **10**(6), e0130019 (2015)
56. L.R. Enders, P. Hur, M.J. Johnson, N.J. Seo, Remote vibrotactile noise improves light touch sensation in stroke survivors' fingertips via stochastic resonance (2013), p. 8
57. T. Kawabe, Inferring sense of agency from the quantitative aspect of action outcome. Conscious. Cogn. **22**(2), 407–412 (2013)
58. D. Leonardis et al., An EMG-controlled robotic hand exoskeleton for bilateral rehabilitation. IEEE Trans. Haptics **8**(2), 140–151 (2015)
59. C. Aruna, A.D. Parameswari, M. Malini, G. Gopu, Voice recognition and touch screen control based wheel chair for paraplegic persons. In *2014 International Conference on Green Computing Communication and Electrical Engineering (ICGCCEE)* (2014), pp. 1–5
60. L.J. Hargrove, K. Englehart, B. Hudgins, A comparison of surface and intramuscular myoelectric signal classification. IEEE Trans. Biomed. Eng. **54**(5), 847–853 (2007)
61. S. Gallagher, The natural philosophy of agency. Philos Compass **2**(2), 347–357 (2007)
62. S. Gallagher, Philosophical conceptions of the self: implications for cognitive science. Trends Cogn. Sci. **4**(1), 14–21 (2000)
63. A.S. Sarıgül, B. Güneri, Some geometric, kinematic, and dynamic considerations on Stewart-Gough platforms with singularity analysis. Robotica **32**(6), 953–966 (2014)
64. C. Fleischer, G. Hommel, A human–exoskeleton interface utilizing electromyography. IEEE Trans. Robot. **24**(4), 872–882 (Aug. 2008)
65. D. Blana, T. Kyriacou, J.M. Lambrecht, E.K. Chadwick, Feasibility of using combined EMG and kinematic signals for prosthesis control: a simulation study using a virtual reality environment. J. Electromyogr. Kinesiol. **29**, 21–27 (Aug. 2016)
66. J.M. Lambrecht, C.L. Pulliam, R.F. Kirsch, Virtual reality environment for simulating tasks with a myoelectric prosthesis: an assessment and training tool. J. Prosthet. Orthot. **23**(2), 89–94 (2011)
67. Z. Tang, K. Zhang, S. Sun, Z. Gao, L. Zhang, Z. Yang, An upper-limb power-assist exoskeleton using proportional myoelectric control. Sensors **14**(4), 6677–6694 (2014)
68. M.A. Powell, R.R. Kaliki, N.V. Thakor, User training for pattern recognition-based myoelectric prostheses: improving phantom limb movement consistency and distinguishability. IEEE Trans. Neural Syst. Rehabil. Eng. **22**(3), 522–532 (2014)
69. C.B. Redd, S.J.M. Bamberg, A wireless sensory feedback device for real-time gait feedback and training. IEEEASME Trans. Mechatron. **17**(3), 425–433 (2012)
70. R. Sigrist, G. Rauter, R. Riener, P. Wolf, Terminal feedback outperforms concurrent visual, auditory, and haptic feedback in learning a complex rowing-type task. J. Mot. Behav. **45**(6), 455–472 (2013)
71. S.M. Radhakrishnan, V. Hatzitaki, A. Vogiannou, D. Tzovaras, The role of visual cues in the acquisition and transfer of a voluntary postural sway task. Gait Posture **32**(4), 650–655 (2010)
72. R. Sigrist, J. Schellenberg, G. Rauter, S. Broggi, R. Riener, P. Wolf, Visual and auditory augmented concurrent feedback in a complex motor task. Presence Teleoperators Virtual Environ. **20**(1), 15–32 (2011)
73. R. Sigrist, G. Rauter, R. Riener, P. Wolf, Augmented visual, auditory, haptic, and multimodal feedback in motor learning: a review. Psychon. Bull. Rev. **20**(1), 21–53 (2013)
74. H.-C. Diener, J. Dichgans, Chapter 22 On the role of vestibular, visual and somatosensory information for dynamic postural control in humans, in *Progress in Brain Research*, ed. by O. Pompeiano, J. H. J. Allum, vol. 76, (Elsevier, New York, 1988), pp. 253–262
75. D. Feygin, M. Keehner, R. Tendick, Haptic guidance: experimental evaluation of a haptic training method for a perceptual motor skill. In *Proceedings 10th Symposium on Haptic Interfaces for Virtual Environment and Teleoperator Systems*. HAPTICS 2002, Orlando, FL, USA (2002), pp. 40–47
76. L. Marchal-Crespo, D.J. Reinkensmeyer, Review of control strategies for robotic movement training after neurologic injury. J. NeuroEng. Rehabil. **6**(1), 20 (2009)

77. E. Rabin, P. DiZio, J. Ventura, J.R. Lackner, Influences of arm proprioception and degrees of freedom on postural control with light touch feedback. J. Neurophysiol. **99**(2), 595–604 (2008)
78. J. Bluteau, S. Coquillart, Y. Payan, E. Gentaz, Haptic guidance improves the visuo-manual tracking of trajectories. PLoS One **3**(3), e1775 (2008)
79. L.A. Jones, Visual and haptic feedback in the control of force. Exp. Brain Res. **130**(2), 269–272 (2000)
80. N. Vuillerme, N. Pinsault, O. Chenu, A. Fleury, Y. Payan, J. Demongeot, Postural destabilization induced by trunk extensor muscles fatigue is suppressed by use of a plantar pressure-based electro-tactile biofeedback. Eur. J. Appl. Physiol. **104**(1), 119–125 (2008)
81. M.R. Afzal, M.-K. Oh, C.-H. Lee, Y.S. Park, J. Yoon, A portable gait asymmetry rehabilitation system for individuals with stroke using a vibrotactile feedback. *BioMed Research International*, 2015. [Online]. Available from https://www.hindawi.com/journals/bmri/2015/375638/. Accessed: 30 Jul 2019
82. M. Afzal, H.-Y. Byun, M.-K. Oh, J. Yoon, Effects of kinesthetic haptic feedback on standing stability of young healthy subjects and stroke patients. J. NeuroEng. Rehabil. **12**(1), 27 (2015)
83. J. Xu et al., Configurable, wearable sensing and vibrotactile feedback system for real-time postural balance and gait training: proof-of-concept. J. NeuroEng. Rehabil. **14**(1), 102 (2017)
84. D.W. Vander Linden, J.H. Cauraugh, T.A. Greene, The effect of frequency of kinetic feedback on learning an isometric force production task in nondisabled subjects. Phys. Ther. **73**(2), 79–87 (1993)
85. D.E. Young, R.A. Schmidt, Augmented kinematic feedback for motor learning. J. Mot. Behav. **24**(3), 261–273 (1992)
86. J.H. Park, C.H. Shea, D.L. Wright, Reduced-frequency concurrent and terminal feedback: a test of the guidance hypothesis. J. Mot. Behav. **32**(3), 287–296 (2000)
87. C.H. Shea, G. Wulf, Enhancing motor learning through external-focus instructions and feedback. Hum. Mov. Sci. **18**(4), 553–571 (1999)
88. S. Abujaber, F. Pozzi, J. Zeni, Influence of weight bearing visual feedback on movement symmetry during sit to stand task. Clin. Biomech. **47**, 110–116 (2017)
89. J.-Y. Chang, G.-L. Chang, C.-J.C. Chien, K.-C. Chung, A.-T. Hsu, Effectiveness of two forms of feedback on training of a joint mobilization skill by using a joint translation simulator. Phys. Ther. **87**(4), 418–430 (2007)
90. Philo Tan Chua et al., Training for physical tasks in virtual environments: Tai Chi. In *IEEE Virtual Reality, 2003. Proceedings*, Los Angeles, CA, USA (2003), pp. 87–94
91. A. Gokeler et al., Feedback techniques to target functional deficits following anterior cruciate ligament reconstruction: implications for motor control and reduction of second injury risk. Sports Med. **43**(11), 1065–1074 (2013)
92. A. Duschau-Wicke, J. von Zitzewitz, A. Caprez, L. Lunenburger, R. Riener, Path control: a method for patient-cooperative robot-aided gait rehabilitation. IEEE Trans. Neural Syst. Rehabil. Eng. **18**(1), 38–48 (2010)
93. A.W. Salmoni, R.A. Schmidt, C.B. Walter, Knowledge of results and motor learning: a review and critical reappraisal. Psychol. Bull. **95**(3), 355–386 (1984)
94. R.A. Schmidt, D.E. Young, S. Swinnen, D.C. Shapiro, Summary knowledge of results for skill acquisition: support for the guidance hypothesis. J. Exp. Psychol. Learn. Mem. Cogn. **15**(2), 352–359 (1989)
95. N.C. Soderstrom, R.A. Bjork, Learning versus performance: an integrative review. Perspect. Psychol. Sci. **10**(2), 176–199 (2015)
96. V. Patoglu, Y. Li, M.K. O'Malley, On the efficacy of haptic guidance schemes for human motor learning, in *World Congress on Medical Physics and Biomedical Engineering, September 7–12, 2009, Munich, Germany*, ed. by O. Dössel, W. C. Schlegel, vol. 25/9, (Springer, Berlin, 2009), pp. 203–206
97. M. Huet, C. Camachon, L. Fernandez, D.M. Jacobs, G. Montagne, Self-controlled concurrent feedback and the education of attention towards perceptual invariants. Hum. Mov. Sci. **28**(4), 450–467 (2009)

98. C.A. Aiken, J.T. Fairbrother, P.G. Post, The effects of self-controlled video feedback on the learning of the basketball set shot. Front. Psychol. **3**, 338 (2012)
99. S. Chiviacowsky, G. Wulf, Self-controlled feedback is effective if it is based on the learner's performance. Res. Q. Exerc. Sport **76**(1), 42–48 (2005)
100. D. Sherwood, Effect of bandwidth knowledge of results on movement consistency. Percept. Mot. Skills **66**, 535–542 (1988)
101. J.E. Goodwin, H.J. Meeuwsen, Using bandwidth knowledge of results to alter relative frequencies during motor skill acquisition. Res. Q. Exerc. Sport **66**(2), 99–104 (1995)
102. A. Badets, Y. Blandin, Observational learning: effects of bandwidth knowledge of results. J. Mot. Behav. **37**(3), 211–216 (2005)
103. C.M. Walsh, S.C. Ling, C.S. Wang, H. Carnahan, enlargethispage*5ptConcurrent versus terminal feedback: it may be better to wait. Acad. Med. J. Assoc. Am. Med. Coll. **84**(10 Suppl), S54–S57 (2009)
104. S. Sülzenbrück, H. Heuer, Type of visual feedback during practice influences the precision of the acquired internal model of a complex visuo-motor transformation. Ergonomics **54**(1), 34–46 (2011)
105. G. Wulf, C.H. Shea, Principles derived from the study of simple skills do not generalize to complex skill learning. Psychon. Bull. Rev. **9**(2), 185–211 (2002)
106. G. Wulf, C.H. Shea, S. Matschiner, Frequent feedback enhances complex motor skill learning. J. Mot. Behav. **30**(2), 180–192 (1998)
107. R. Sigrist, G. Rauter, L. Marchal-Crespo, R. Riener, P. Wolf, Sonification and haptic feedback in addition to visual feedback enhances complex motor task learning. Exp. Brain Res. **233**(3), 909–925 (2015)
108. J. Sadowski, A. Mastalerz, T. Niznikowski, Benefits of bandwidth feedback in learning a complex gymnastic skill. J. Hum. Kinet. **37**(1), 183–193 (2013)
109. R.J. Triolo, C. Bieri, J. Uhlir, R. Kobetic, A. Scheiner, E.B. Marsolais, Implanted functional neuromuscular stimulation systems for individuals with cervical spinal cord injuries: clinical case reports. Arch. Phys. Med. Rehabil. **77**(11), 1119–1128 (1996)
110. M. Schiefer, D. Tan, S.M. Sidek, D.J. Tyler, Sensory feedback by peripheral nerve stimulation improves task performance in individuals with upper limb loss using a myoelectric prosthesis. J. Neural Eng. **13**(1), 016001 (2016)
111. T.A. Kuiken et al., Targeted reinnervation for enhanced prosthetic arm function in a woman with a proximal amputation: a case study. Lancet Lond. Engl. **369**(9559), 371–380 (2007)
112. P.D. Marasco et al., Illusory movement perception improves motor control for prosthetic hands. *Sci. Transl. Med.***10**(432), eaao6990 (2018)
113. S. Raspopovic et al., Restoring natural sensory feedback in real-time bidirectional hand prostheses. Sci. Transl. Med. **6**(222), 222ra19 (2014)
114. A. Khasnobish, M. Pal, D. Sardar, D.N. Tibarewala, A. Konar, Vibrotactile feedback for conveying object shape information as perceived by artificial sensing of robotic arm. Cogn. Neurodyn. **10**(4), 327–338 (2016)
115. K. Grechuta, J. Guga, G. Maffei, B. Rubio Ballester, P.F.M.J. Verschure, Visuotactile integration modulates motor performance in a perceptual decision-making task. Sci. Rep. **7**(1), 3333 (2017)
116. K. Grechuta, L. Ulysse, B.R. Ballester, P.F.M.J. Verschure, Self beyond the body: task-relevant distal cues modulate performance and body ownership, Neuroscience, preprint (2018)
117. H.H. Ehrsson, That's my hand! activity in premotor cortex reflects feeling of ownership of a limb. Science **305**(5685), 875–877 (2004)
118. H.H. Ehrsson, Touching a rubber hand: feeling of body ownership is associated with activity in multisensory brain areas. J. Neurosci. **25**(45), 10564–10573 (2005)
119. A. Tapal, E. Oren, R. Dar, B. Eitam, The sense of agency scale: a measure of consciously perceived control over one's mind, body, and the immediate environment. Front. Psychol. **8**, 1552 (2017)

120. C. Jeunet, L. Albert, F. Argelaguet, A. Lecuyer, 'Do you feel in control?': towards novel approaches to characterise, manipulate and measure the sense of agency in virtual environments. IEEE Trans. Vis. Comput. Graph. **24**(4), 1486–1495 (2018)
121. C.Z.-H. Ma, W.C.-C. Lee, A wearable vibrotactile biofeedback system improves balance control of healthy young adults following perturbations from quiet stance. Hum. Mov. Sci. **55**, 54–60 (2017)
122. K.H. Sienko, M. Balkwill, L.I.E. Oddsson, C. Wall, The effect of vibrotactile feedback on postural sway during locomotor activities. J. NeuroEng. Rehabil. **10**(1), 93 (2013)
123. H.J.B. Witteveen, L. de Rond, J.S. Rietman, P.H. Veltink, Hand-opening feedback for myoelectric forearm prostheses: performance in virtual grasping tasks influenced by different levels of distraction. J. Rehabil. Res. Dev. **49**(10), 1517 (2012)
124. W. Nanhoe-Mahabier, J.H. Allum, E.P. Pasman, S. Overeem, B.R. Bloem, The effects of vibrotactile biofeedback training on trunk sway in Parkinson's disease patients. Parkinsonism Relat. Disord. **18**(9), 1017–1021 (2012)
125. E. Raveh, S. Portnoy, J. Friedman, Adding vibrotactile feedback to a myoelectric-controlled hand improves performance when online visual feedback is disturbed. Hum. Mov. Sci. **58**, 32–40 (2018)
126. J.D. Brown et al., An exploration of grip force regulation with a low-impedance myoelectric prosthesis featuring referred haptic feedback. J. NeuroEng. Rehabil. **12**, 104 (2015)
127. T.J. Carroll, S. Riek, R.G. Carson, Neural adaptations to resistance training: implications for movement control. Sports Med. Auckl. NZ **31**(12), 829–840 (2001)
128. P. Stegall, D. Zanotto, S.K. Agrawal, Variable damping force tunnel for gait training using ALEX III. IEEE Robot. Autom. Lett. **2**(3), 1495–1501 (2017)
129. L. Hargrove, Y. Losier, B. Lock, K. Englehart, B. Hudgins, A real-time pattern recognition based myoelectric control usability study implemented in a virtual environment. In *2007 29th Annual International Conference of the IEEE Engineering in Medicine and Biology Society*, Lyon, France (2007), pp. 4842–4845
130. J.-U. Chu, I. Moon, Y.-J. Lee, S.-K. Kim, M.-S. Mun, A supervised feature-projection-based real-time emg pattern recognition for multifunction myoelectric hand control. IEEEASME Trans. Mechatron. **12**(3), 282–290 (2007)
131. Y. Geng et al., A robust sparse representation based pattern recognition approach for myoelectric control. IEEE Access **6**, 38326–38335 (2018)
132. L.F. Engels, A.W. Shehata, E.J. Scheme, J.W. Sensinger, C. Cipriani, When less is more – discrete tactile feedback dominates continuous audio biofeedback in the integrated percept while controlling a myoelectric prosthetic hand. Front. Neurosci. **13**, 578 (2019)
133. S.M. Wurth, L.J. Hargrove, A real-time comparison between direct control, sequential pattern recognition control and simultaneous pattern recognition control using a Fitts' law style assessment procedure. J. NeuroEng. Rehabil. **11**(1), 91 (2014)
134. R. Chereshnev, A. Kertész-Farkas, GaIn: human gait inference for lower limbic prostheses for patients suffering from double trans-femoral amputation. Sensors **18**(12), 4146 (2018)
135. A. Fleming, S. Huang, H. Huang, Proportional myoelectric control of a virtual inverted pendulum using residual antagonistic muscles: toward voluntary postural control. IEEE Trans. Neural Syst. Rehabil. Eng. **27**(7), 1473–1482 (2019)
136. O.W. Samuel, et al., A novel time-domain descriptor for improved prediction of upper limb movement intent in EMG-PR system. In *2018 40th Annual International Conference of the IEEE Engineering in Medicine and Biology Society (*EMBC*)*, Honolulu, HI (2018), pp. 3513–3516
137. F. Leone et al., Simultaneous sEMG classification of hand/wrist gestures and forces. Front. Neurorobot. **13**, 42 (2019)
138. R. Nataraj, M.L. Audu, R.F. Kirsch, R.J. Triolo, Comprehensive joint feedback control for standing by functional neuromuscular stimulation—a simulation study. IEEE Trans. Neural Syst. Rehabil. Eng. **18**(6), 646–657 (2010)
139. R. Nataraj, M.L. Audu, R.J. Triolo, Comparing joint kinematics and center of mass acceleration as feedback for control of standing balance by functional neuromuscular stimulation. J.

NeuroEng. Rehabil. **9**(1), 25 (2012)
140. R. Nataraj, M.L. Audu, R.J. Triolo, Center of mass acceleration feedback control of standing balance by functional neuromuscular stimulation against external postural perturbations. IEEE Trans. Biomed. Eng. **60**(1), 10–19 (2013)
141. A.B. Ajiboye, R.F.H. Weir, A heuristic fuzzy logic approach to EMG pattern recognition for multifunctional prosthesis control. IEEE Trans. Neural Syst. Rehabil. Eng. **13**(3), 280–291 (2005)
142. B. Crawford, Real-time classification of electromyographic signals for robotic control, p. 6
143. M.V. Liarokapis, P.K. Artemiadis, P.T. Katsiaris, K.J. Kyriakopoulos, E.S. Manolakos, Learning human reach-to-grasp strategies: towards EMG-based control of robotic arm-hand systems. In *2012 IEEE International Conference on Robotics and Automation*, St Paul, MN, USA (2012), pp. 2287–2292
144. D.T. Westwick, E.A. Pohlmeyer, S.A. Solla, L.E. Miller, E.J. Perreault, Identification of multiple-input systems with highly coupled inputs: application to emg prediction from multiple intracortical electrodes. Neural Comput. **18**(2), 329–355 (2006)
145. R.E. Johnson, K.P. Kording, L.J. Hargrove, J.W. Sensinger, Adaptation to random and systematic errors: comparison of amputee and non-amputee control interfaces with varying levels of process noise. PLoS One **12**(3), e0170473 (2017)
146. H.J. Chiel, R.D. Beer, R.D. Quinn, K.S. Espenschied, Robustness of a distributed neural network controller for locomotion in a hexapod robot. IEEE Trans. Robot. Autom. **8**(3), 293–303 (1992)
147. E.A. Clancy, N. Hogan, Multiple site electromyograph amplitude estimation. IEEE Trans. Biomed. Eng. **42**(2), 203–211 (1995)
148. S. Zhou, K. Yin, Z. Liu, F. Fei, J. Guo, sEMG-based hand motion recognition by means of multi-class adaboost algorithm. In *2017 IEEE International Conference on Robotics and Biomimetics (*ROBIO*)*, Macau (2017), pp. 1056–1061
149. K. Walsh, R. Nataraj, Simulated nervous system lesions in machine learning for myoelectric control, Biomedical Engineering, Stevens Institute of Technology (2019)

Index

R. Vinjamuri (ed.), *Advances in Motor Neuroprostheses*,
https://doi.org/10.1007/978-3-030-38740-2

H

I

T

The manufacturer's authorised representative in the EU is Springer Nature Customer Service Centre GmbH, Europaplatz 3, 69115 Heidelberg, Germany. If you have any concerns regarding our products, please contact ProductSafety@springernature.com

Printed and bound by CPI Group (UK) Ltd, Croydon, CR0 4YY

30/09/2026

02225574-0014